Manuel Feuchter

Investigations on Joule heating applications by multiphysical continuum simulations in nanoscale systems

**Schriftenreihe
des Instituts für Angewandte Materialien**
Band 43

Karlsruher Institut für Technologie (KIT)
Institut für Angewandte Materialien (IAM)

Eine Übersicht aller bisher in dieser Schriftenreihe erschienenen Bände
finden Sie am Ende des Buches.

Investigations on Joule heating applications by multiphysical continuum simulations in nanoscale systems

by
Manuel Feuchter

Dissertation, Karlsruher Institut für Technologie (KIT)
Fakultät für Maschinenbau
Tag der mündlichen Prüfung: 08.07.2014

Impressum

Scientific
Publishing

Karlsruher Institut für Technologie (KIT)
KIT Scientific Publishing
Straße am Forum 2
D-76131 Karlsruhe

KIT Scientific Publishing is a registered trademark of Karlsruhe
Institute of Technology. Reprint using the book cover is not allowed.

www.ksp.kit.edu

Print on Demand 2014

ISSN 2192-9963
ISBN 978-3-7315-0261-6
DOI   10.5445/KSP/1000042982

Investigations on Joule heating applications by multiphysical continuum simulations in nanoscale systems

Zur Erlangung des akademischen Grades

Doktor der Ingenieurwissenschaften

der Fakultät für Maschinenbau

Karlsruher Institut für Technologie (KIT)

genehmigte

DISSERTATION

von

Dipl.–Ing. Manuel Klaus Ludwig Feuchter

geboren am 20.11.1983

in Wertheim

Tag der mündlichen Prüfung:	08.07.2014
Hauptreferent:	Prof. Dr.–Ing. M. Kamlah
Korreferent:	Prof. Dr. rer. nat. C. Jooss
Korreferent:	Prof. Dr. rer. nat. O. Kraft

'No subject has more extensive relations with
the progress of industry and the natural sciences;
for the action of heat is always present,
it influences the processes of the arts,
and occurs in all the phenomena of the universe.'
—

Jean Baptiste Joseph Fourier [cf. 118, p. 241]

Abstract

A requirement for future thermoelectric applications are poor heat conducting materials. Nowadays, several nanoscale approaches are used to decrease the thermal conductivity of this material class. A promising approach applies thin multilayer structures, composed of known materials. Here, the nanoscale stacking affects the heat conduction and provokes new emergent thermal properties; However, the heat conduction through these materials is not well understood yet. Therefore, a reliable measurement technique is required to measure and understand the heat propagation through these materials. In this work, the so-called 3ω-method is focused upon to investigate thin strontium titanate (STO) and praseodymium calcium manganite (PCMO) layer materials. Previously unexamined macroscopic influence factors within a 3ω-measurement are considered in this thesis by Finite Element simulations. Thus, this work furthers the overall understanding of a 3ω-measurement, and allows precise thermal conductivity determinations. Moreover, new measuring configurations are developed to determine isotropic and anisotropic thermal conductivities of samples from the micro- to nanoscale. Since no analytic solutions are available for these configurations, a new evaluation methodology is presented to determine emergent thermal conductivities by Finite Element simulations and Neural Networks.

Kurzfassung

Zukünftige thermoelektrische Anwendungen erfordern thermisch schlecht leitende Materialien. Hierfür werden heutzutage verschiedene Ansätze verfolgt um die Wärmeleitfähigkeitszahl dieser Materialklasse zu verringern. Ein vielversprechender Ansatz verwendet dünne Vielschichtstrukturen die sich aus bekannten Materialien zusammen setzen. Hier wird durch nanoskaliges Schichten die Wärmeleitung beeinflußt und es werden neue thermische Eigenschaften hervorgerufen. Jedoch ist die Wärmeleitung durch solch ein Material bis heute noch nicht bis ins Detail verstanden. Zu diesem Zweck bedarf es einer verläßlichen Messmethode, um die Wärmeausbreitung durch solch ein Material messen und somit verstehen zu können. Diese Arbeit konzentriert sich auf die so genannte 3ω-Methode zur Erforschung dünner Strontium Titanat (STO) und Praseodym Calcium Manganit (PCMO) Schichtmaterialien. Bisher unbeachtete makroskopische Einflußfaktoren, die während einer 3ω-Messung auftreten, werden in dieser Arbeit durch Finite Element Simulationen berücksichtigt. Dadurch trägt dieses Werk zum Gesamtverständnis einer 3ω-Messung bei und erlaubt eine präzise Bestimmung der Wärmeleitfähigkeitszahl. Darüber hinaus werden neue Messkonfigurationen zur Bestimmung der isotropen als auch anisotropen Wärmeleitfähigkeitszahl für mikro- bis nanoskalige Proben entwickelt. Da für diese Messkonfigurationen keine analytischen Lösungen verfügbar sind, wird eine neue Methodik zur Auswertung und Bestimmung der Wärmeleitfähigkeitszahl vorgestellt, die Finite Element Simulationen und Neuronale Netze kombiniert.

Contents

Nomenclature

Greek Symbols

α	linear temperature coefficient	[1/K]
β	linear thermal expansion coefficient	[1/K]
Ω_H	area of the heater	
$\Omega_{lay/m}$	area of the layer or material	
Ω_s	area of the substrate	
σ_s	stress tensor	
Υ	amplitude of motion	
ε	infinitesimal strain tensor	
ε_T	thermal strain	
χ, ω, ψ, ϕ	complex terms of Borca-Tasciuc's solution	
ΔR	electric resistance oscillation	[Ω]
ΔT_H^B	temperature amplitude due to Borca-Tasciuc including heater properties	[K]
ΔT_h^B	temperature amplitude due to Borca-Tasciuc without heater properties	[K]
ΔT_h^C	temperature amplitude due to Cahill	[K]
ΔT_{lay}^C	temperature amplitude due to layer	[K]
$\Delta \check{T}$	general temperature oscillation	[K]
ΔT	temperature amplitude in the heater/material	[K]
ϵ	permittivity	[As/Vm]
ϵ_0	vacuum constant permittivity	[As/Vm]
ϵ_r	relative permittivity of specific material	[$-$]
η	Neural Network gradient scaling parameter	

Γ	boundary for the partial differential equation	
γ	Euler-Maschoneri constant	$[-]$
$\Gamma_{\mathbb{H}_{1-4}}$	boundary for the magnetic field	
Γ_{Int}	interface boundary	
κ_{eff}	effective thermal conductivity	$[\,W/mK\,]$
κ	thermal conductivity	$[\,W/mK\,]$
κ_e	thermal conductivity due to electrons	$[\,W/mK\,]$
κ_H	thermal conductivity of the heater	$[\,W/mK\,]$
$\kappa_k,\,\kappa_{k+1}$	thermal conductivities of two contacting materials	$[\,W/mK\,]$
κ_{lay}	thermal conductivity of the layer	$[\,W/mK\,]$
$\kappa_{m_{\text{crs}}}$	cross-plane thermal conductivity of the investigated material	$[\,W/mK\,]$
$\kappa_{m_{\text{in}}}$	in-plane thermal conductivity of the investigated material	$[\,W/mK\,]$
κ_m	thermal conductivity of the investigated material	$[\,W/mK\,]$
κ_{ph}	thermal conductivity due to phonons	$[\,W/mK\,]$
κ_s	thermal conductivity of the substrate	$[\,W/mK\,]$
$\mathcal{E}(\mathcal{W})$	Neural Network regulation term	
$\mathcal{G}(\mathcal{W})$	Neural Network error term	
Q	Neural Network activation function	
$\mathcal{W}^{\mathcal{T}+1}$	Neural Network iterative synaptic weight	
$\mathcal{W}^{\mathcal{T}}$	Neural Network actual synaptic weight	
$\mathcal{W}_{\text{ij}}$	Neural Network synaptic weights	
$\mathcal{X}_{\text{j}}$	Neural Network input vector	
$\mathcal{Y}_{\text{j}}$	Neural Network output vector	
μ	permeability	$[\,Vs/Am\,]$
μ_0	vacuum constant permeability	$[\,Vs/Am\,]$

μ_r	relative permeability of specific material	$[-]$
ν	Poisson's ratio	$[-]$
Ω	area of the partial differential equation	
ω	angular frequency of the current	$[\,1/s\,]$
Π	Peltier coefficient	$[\,W/A\,]$
ρ	mass density	$[\,kg/m^3\,]$
ρ_e	electric charge density	$[\,C/m^2\,]$
ρ_m	mass density of the investigated material	$[\,kg/m^3\,]$
σ	mechanical stress	$[\,N/m^2\,]$
σ_b	Boltzmann constant	$[\,J/K\,]$
σ_{ec}	electrical conductivity	$[\,1/\Omega m\,]$
τ	mechanical shear stress	$[\,N/m^2\,]$
Θ	correction value due to boundary mismatch	$[-]$
θ	angular of rotation	$[°]$
φ	contact potential	
ϱ_0	specific electric resistivity	$[\,\Omega m\,]$
$\varrho_0(T)$	temperature dependent specific electric resistivity	$[\,\Omega m\,]$
ς	phase shift	$[°]$
$\wp_e$	electric charge density, complex quantity	

Latin Symbols

$1/q$	thermal penetration depth	$[\,m\,]$
$\mathbf{f}$	volume force	
$\mathbb{A}_{j_{init}}$	magnetic vector potential, initial complex quantity	
$\mathbb{A}_j$	magnetic vector potential, complex quantity	$[\,Vs/m\,]$
$\mathbb{B}$	magnetic flux density, complex quantity	
$\mathbb{D}$	electric flux density, complex quantity	

$\mathbb{E}$	electric field intensity, complex quantity	
$\mathbb{H}$	magnetic field intensity, complex quantity	
$\mathbb{H}_{\mathrm{mat}}$	magnetic field intensity in a certain material	
$\mathbb{J}$	electric current density, complex quantity	
$\mathcal{F}$	Fourier cosine transform	
$\mathbf{B}$	magnetic flux density	$[\,\mathrm{Vs/m^2}\,]$
$\mathbf{C}$	elasticity tensor	
$\mathbf{D}$	electric flux density	$[\,\mathrm{As/m^2}\,]$
$\mathbf{E}$	electric field intensity	$[\,\mathrm{V/m}\,]$
$\mathbf{H}$	magnetic field intensity	$[\,\mathrm{A/m}\,]$
$\mathbf{j}$	electric current density	$[\,\mathrm{A/m^2}\,]$
$\mathbf{u}$	mechanical displacement vector	
A	cross-sectional area of the heater	$[\,\mathrm{m^2}\,]$
a	height of the heater	$[\,\mathrm{m}\,]$
A_p	circular surface area of the pillar	$[\,\mathrm{m^2}\,]$
b	half heater width	$[\,\mathrm{m}\,]$
b_{eff}	effective half heater width	$[\,\mathrm{m}\,]$
c_{P_m}	specific heat capacity of the investigated material	$[\,\mathrm{J/kgK}\,]$
c_P	specific heat capacity	$[\,\mathrm{J/kgK}\,]$
dur_p	pulse duration	$[\,\mathrm{s}\,]$
dT/dR	specific temperature to resistance behavior	$[\,\mathrm{K/\Omega}\,]$
d_{lay}	thickness of the layer	$[\,\mathrm{m}\,]$
d_{PCMO}	thickness of pillar PCMO layer	$[\,\mathrm{m}\,]$
D_s	thermal diffusivity of the substrate	$[\,\mathrm{m^2/s}\,]$
d_s	thickness of the substrate	$[\,\mathrm{m}\,]$
e	surface emissivity	$[\,-\,]$
fdp	regime length	$[\,\mathrm{s}\,]$

f_c	frequency of the current	[Hz]
f_p	pulse repetition rate	[s]
G	shear modulus	[N/m^2]
htc	heat transfer coefficient	[W/m^{2}K]
i	complex number	[–]
I_0	peak current	[A]
I_0^B	modified Bessel function of first kind and zero order	
j_0	peak current density	[A/m^2]
K_0^B	modified Bessel function of second kind and zero order	
L	length of the heater	[m]
$L_h(z)$	length of the heater in z-direction	[m]
n	normal to the cross-sectional cut	
P	released power	[W]
p	source term	[W/m^3]
P_0	applied power amplitude	[W]
P_0/L	power per length	[W/m]
Q	amount of heat	[W]
q	heat flux	[W/m^2]
$R(T)$	measured electric resistance of the PCMO pillar	[Ω]
R	electric resistance	[Ω]
r	distance form line source	[m]
R_0	average electric resistance	[Ω]
r_p	radius of the pillar	[m]
sdp	step size	[s]
t	time	[s]
T_0	ambient temperature	[K]
T_{av}	average temperature rise	[K]
T_{co}	constant temperature rise	[K]

T_H	temperature in the heater	[K]
T_{init}	initial temperature	[K]
T_k, T_{k+1}	temperatures	
	of two contacting materials	[K]
$T_{mean_{max}}$	maximum mean temperature in the heater	[K]
$T_{mean_{min}}$	minimum mean temperature in the heater	[K]
$T_{mean}(t)$	time dependent mean temperature in the heater	[K]
T_m	temperature in the investigated material	[K]
T_s	temperature in the substrate	[K]
U	voltage	
$U_{3\omega}$	third harmonic voltage	[V]
$u_{x,y,z}$	mechanical displacement in the respective spatial	
	direction	[m]
x, y, z	spatial coordinates	[m]
E_σ	Young's modulus	[N/m^2]
f	arbitrary function	
k	complex quantity of Helmholtz-equation	
S	surface area	[m^2]
I	electric current	[A]
S	Seebeck coefficient	[μV/K]
T	temperature	[K]
ZT	figure of merit for thermoelectrics	[−]

Miscellaneous

FES	Finite Element Simulation
SEM	scanning electron microscopy
SIN	Si_3N_4

BC   boundary condition
PCMO $Pr_{1-x}Ca_xMnO_3$
RRAM   resistive random access memory
STO $SrTiO_3$
YSZ   compound of ZrO_2 and Y_2O_3
PMMA   Polymethyl Methacrylate $C_5H_8O_2$

Operators

$\nabla \cdot$ divergence
$\nabla \times$ rotation
∇ gradient

1 Introduction

Motivation

By the year 2035, the global energy demand will have increased by one-third of today's consumption [75]. Although it is controversially discussed from when on primary used energy resources will run short, it is fact that costs for energy sources increased significantly in the recent years and probably will in future [143]. To overcome this dilemma, new technologies for energy production are necessary, accompanied by efficiently using energy resources. In many cases, heat engines are used to convert primary energy sources[1] into mechanic and electric energy. However, a tremendous amount of energy is lost by waste heat [cf. 158]. Here, thermoelectric generators can contribute to increase the overall efficiency by turning waste heat into electricity. A typical application example are conventionally driven cars Fig.1(a). Hot exhaust gases pass off into the environment without any use. Here, the car's overall efficiency could benefit from waste heat recovery.

However, besides the enhancement of existing heat engines, thermoelectric generators can be used also at smaller scales in mini devices for energy harvesting. Conceivable applications are embedded electronic devices into human clothing such as sensors, mobile phones or media players Fig.1(b). Application areas that don't effect daily life of general public are special stand-alone energy systems. Facing the challenge of constant energy support over several years, some satellites Fig.1(c) use radioisotope thermoelectric generators since the early 1960s [129]. This kind of generator is also used in the famous self-sustaining space rover Curiosity on Mars Fig.1(d).

---

[1]e.g. oil, coal and gas

Based on the working principle, thermoelectric converters can also be used to cool devices. In recent years, a trend towards electro mobility emerged. The desired operating distance and the apparent mass of the vehicle requires high performance batteries. These batteries emit a significant amount of heat in comparison to small electronic devices when discharged. Here, a cooling regulation might be needed while the car is operated Fig.1(e). Less spectacular, but prevalent are portable refrigerators.

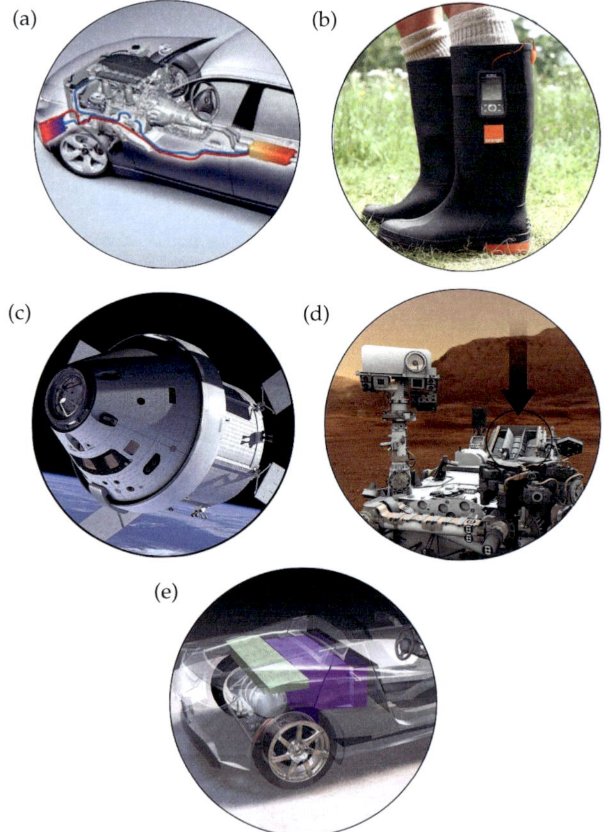

Figure 1: Examples for thermoelectric generators: (a) application at an exhaust [65]; (b) mini devices [153]; (c) stand-alone satellites [135]; (d) self-sustaining space rovers [119]; (e) battery cooling [55].

However, the yield of energy conversion in these materials is still quite low today [cf. 122, 158]. Moreover, the materials with the highest thermoelectric efficiency are in most cases not environmentally friendly and relatively expensive [98].   These facts limit the thermoelectric energy conversion to niche applications. Therefore, it is desirable to develop new, environmentally friendly materials with a good cost-benefit ratio to open up this technology for general public applications. This can be achieved either by cheap and abundant materials or by attaining higher efficiency.

Although, thermoelectric materials have been in use for decades for energy conversion, their efficiency has remained at the same level for almost half a century. This was due to a lack of physical insights and new materials. Finally, a new conceptual approach arised in 1993. Hicks and Dresselhaus [70] reported on quantum size effects on the thermoelectric efficiency. Since then, nanostructuring[2] has been applied on nanowires [72], phononic nanomesh structures [168], quantum-dot systems [25], nanograined bulk materials [131] and multilayer structures [157] to enhance the thermoelectric efficiency [cf. 122].

Together with material physicist groups around Blöchl[3], Jooss[4] and Volkert[5], we are bound into a priority program of the German research association (DFG SPP 1386). This program was founded to enhance thermoelectric efficiency for power generation from heat through nanostructured materials. As part of this program, we focus on nanoscale structured multilayer and superlattice systems.

---

[2]Further information about nanoscale thermoelectrics is given by Pichanusakorn [129].
[3]Institute of Theoretical Physics, Clausthal University of Technology
[4]Institute of Materials Physics, University of Göttingen
[5]Institute of Materials Physics, University of Göttingen

The high potential for multilayer thermoelectric materials has been shown by Venkatasubramanian et al. [156] in 2000. They increased the thermoelectric efficiency significantly by suppressing the thermal conductivity perpendicular to the layered materials.[6] Multilayers are amorphous or polycrystalline layered materials. In contrast, in superlattice systems each layer is single crystalline [31]. However, both consist of two different materials, which alternate in a stack. The thickness of each layer material is on the nanometer scale. For such systems the thermal conductivity of the stack cannot simply be predicted from each layer material, because here the transport of thermal energy depends on the phonon propagation across the interfaces and along surfaces at various wavelengths. Hence, the whole package represents a material with new resultant thermal conductivity. Consequently, Fourier's law of heat conduction [cf. Eq.1.7] would be represented with an effective thermal conductivity κ_{eff}.

However, to design highly efficient layered thermoelectric materials, it is necessary to understand the influence of specific and intrinsic sample properties onto the anisotropic thermal transport of each layer material and the stack. Therefore, a reliable measurement technique is required to determine the cross-plane and in-plane thermal conductivity of nanoscale samples.

---

[6]Although the absolute values are discussed controversially in literature, it indicates a tremendous decrease of the thermal conductivity.

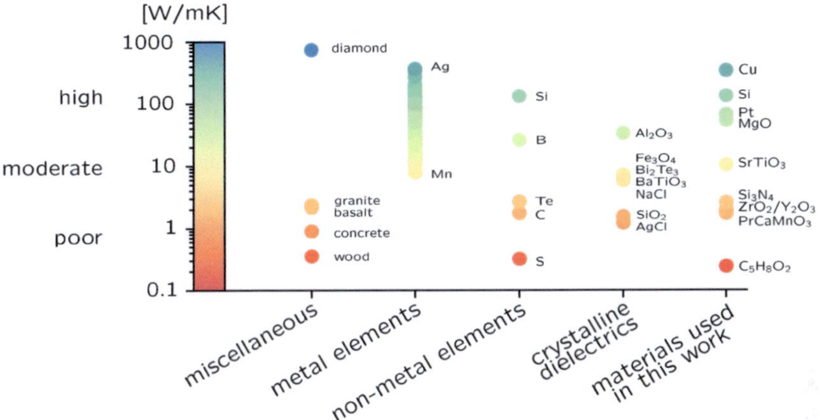

Figure 2: Exemplary collection of thermal conductivity values of different solid materials at room temperature [cf. 71, 106].

Objectives of this work

In this work, we mainly focus on the 3ω-method to determine the thermal conductivities of the investigated materials. While several techniques exist to determine the thermal conductivity of solids, thin films or multilayers, the 3ω-method is one of the most well-established due to its high accuracy [78, 126]. This method is especially convenient for poor heat conductors [28], such as thermoelectric materials. In this work, we distinguish the thermal conductivity of the materials to be either poorly-conductive, moderately-conductive or highly-conductive. Since thermoelectric materials are applied with direct currents and relative constant temperature gradients, these terms refer only to the materials thermal conductivity throughout this work. In Fig.2, an overview is given for the thermal conductivity of different solid materials.

However, there is still a demand to further the overall understanding of macroscopic influence factors within a 3ω-measurement in order to determine accurate thermal conductivities. Moreover, no analytic solutions are available for complex measuring configurations, which could allow to measure the anisotropic thermal conductivity in poorly-conductive materials.

Therefore, the objectives of this work are: Examine macroscopic influence factors within a 3ω-measurement, develop new geometry configurations to measure the cross-plane and in-plane thermal conductivity, and establish a new methodology, combining experiments and simulations, to identify the isotropic and anisotropic thermal conductivity of materials with nanoscale thickness, such as layered films and pillar geometries.

Outline of the present work

In the following introduction chapter, the fundamentals of thermoelectricity are introduced and the materials of major interest are presented. Subsequently, the notion of macroscopic heat conduction in a continuum is derived. Chapter 2 covers the concept and working principle of the 3ω-method. Different analytic solutions of various geometry configurations are compared, the limits are examined and discussed. In chapter 3, the principles of the Finite Element Model are presented including the governing equations, the transient analysis and information about the meshes. In chapter 4, different geometry structures are investigated. First, different macroscopic influence factors are studied for heat sources on top of bulk material configurations and the relevance for a 3ω-measurement is examined. Second, classic monolayer and multilayer configurations are considered. Third, bottom electrode geometries are investigated for bulk-like to thin layer materials. Fourth, two-dimensional and three-dimensional membrane structures are studied. At last, pillar structures are investigated for current induced pulse heating applications. In chapter 5, a new methodology is presented to determine the anisotropic thermal conductivity with Finite Element simulations and Neural Networks. Chapter 6.1 contains two application examples for the thermal conductivity determination with the new presented method. Finally, this work is summarized in chapter 7.

1.1 Thermoelectricity and material selection

Thermoelectricity

In general, three distinct thermoelectric effects exist. The first thermoelectric effect was discovered by Thomas Johann Seebeck in 1820-1821, however he dated his memorandum 1822-1823 [142]. The 'Seebeck-effect' induces a voltage when two electrically conducting materials in a circuit (thermocouple) have different temperatures (T_1, T_2) at their junctions [Fig.3].

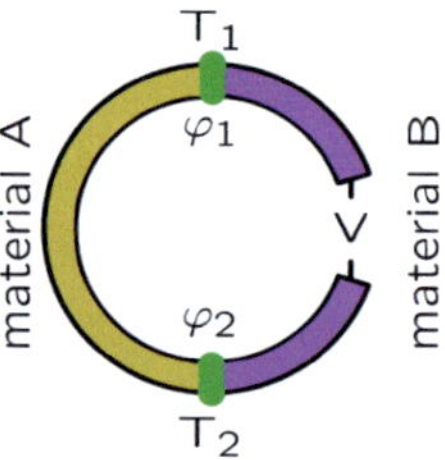

Figure 3: 'Seebeck-effect': Induced voltage in a thermocouple [cf. 10, 137].

The thermo voltage ΔV between the material junctions is defined as the difference of the contact potentials $\varphi_{1/2}$ [10] or the Seebeck coefficient S times the temperature difference [74, p. 100]

$$\Delta V = \varphi_1 - \varphi_2 = S \cdot (T_1 - T_2) \,. \tag{1.1}$$

Shortly afterwards, Jean Charles Athanase Peltier (1834) discovered the reversion of the 'Seebeck-effect'. Applying a direct current through a thermocouple, consisting of two different materials, leads one intersection to heat up and the other one to cool down [Fig.4] [74, p. 100].[7]

---

[7]The heat produced by the 'Peltier-effect' at the up heating intersection surpasses by far

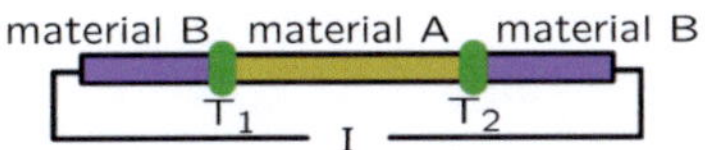

Figure 4: 'Peltier-effect': Electric current flow in a thermocouple [cf. 74].

The amount of heat per unit time Q, which is absorbed at one intersection and liberated at the other intersection is defined as [138]

$$Q = \Pi \cdot I. \tag{1.2}$$

Here, Π is the Peltier coefficient and I is the electric current. The 'Peltier-effect' is reversible and describes the change in heat content at an intersection between two different materials. The change in heat content results from the flow of electric current across it [137]. The direction of the current flow determines whether the intersection heats up or cools down.[8] A connection between the Seebeck coefficient and the Peltier coefficient is given [137] via the temperature by

$$\Pi = T \cdot S. \tag{1.3}$$

Approximately two decades after the 'Peltier-effect' was discovered, William Thomson (1848-1854) observed an additional thermal effect due to the electric current [52]. The 'Thomson-effect' describes the fact that heat is either liberated or absorbed within the leg of a thermocouple if an electric current flows through, while a temperature gradient exists.

---

the heat produced by the 'Joule-effect' in the material [63, p. 349].

[8]Joule heating takes place in every electrical conductor when an electrical current flows through it and does not require the existence of two different materials. Furthermore, Joule heating is independent of the current's direction.

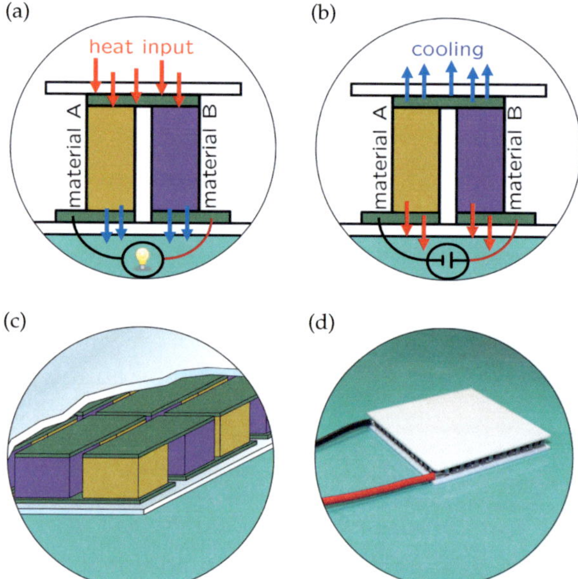

Figure 5: Working principle of a thermocouple for (a) electric energy generation and (b) cooling application. Fig. (c) shows the arrangement of multiple thermocouples in a module [cf. 145] and Fig. (d) shows a fabricated thermoelectric module [150].

In thermoelectric converters, the thermocouple is rearranged to obtain larger intersections between material A and B [Fig.5(a) and 5(b)]. Thus, the intersections are better exposed for thermal contacts. A thermocouple can be used in two ways. First, applying an external heat input induces a thermo voltage by the 'Seebeck-effect'. Thus waste heat can be transferred to electric energy. Second, applying an electric current results in cooling of one side of the thermocouple (absorption of heat) and heating up the other side of the thermocouple (heat rejection). Hence, electric energy can be used by the 'Peltier-effect' for cooling applications. Since both conversions involve a temperature gradient, the 'Thomson-effect' occurs for both applications of the thermoelectric converter, energy harvesting by the 'Seebeck-effect' and cooling application by the 'Peltier-effect'. However, for energy conversion in thermoelectric generators, the 'Thomson-effect' is not of primary importance [137].

A key property of the converting modules is the material's efficiency, described by the 'figure of merit for thermoelectrics' (ZT):

$$\mathrm{ZT} = \frac{\sigma_{ec} \cdot \mathrm{S}^2}{\kappa} \cdot \mathrm{T} \quad ; \quad \kappa = \kappa_e + \kappa_{ph} \,. \tag{1.4}$$

Here, σ_{ec} is the electrical conductivity, S the Seebeck coefficient, T the temperature and κ the thermal conductivity. The thermal conductivity is composed of a part due to electrons κ_e and a part due to phonons κ_{ph} [145]. Since every thermoelectric material has a peak performance at a certain temperature [cf. 145], the temperature is included in the 'figure of merit' [137]. In general, this peak performance is in the range of $370 - 1270$ K [cf. 145]. Nowadays, industrial applications use thermoelectric materials of which the maximum $\mathrm{ZT} \approx 1$ [158].

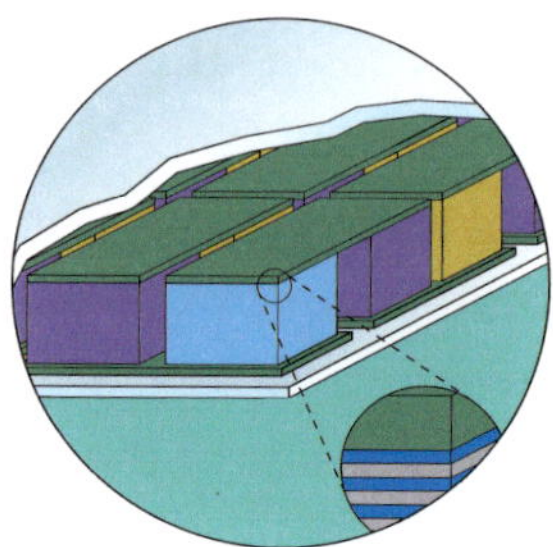

Figure 6: Replacement of bulk-like material by multilayer structures.

Material selection

Consequently, the choice of materials is essential for efficient thermoelectric converters. The conflicting terms of the 'figure of merit' need to be optimized in order to increase the ZT value [cf. 145]. However, the distinct terms of the 'figure of merit' are connected with each other since all quantities depend on electron properties. Here, the new conceptual approach of Hicks and Dresselhaus [70] could allow the reduction of the phononic part of the thermal conductivity κ_{ph} by structuring at the nanoscale and thus decrease the overall thermal conductivity. Normally, each leg of a thermocouple consists of bulk-like materials. These could be replaced by multilayer structures [cf. Fig.6]. For the development of sustainable thermoelectrics, we focus on environmentally friendly oxide materials. The two following materials are of major interest in this work.[9]

The first material we focus on is the perovskite oxide material strontium titanate $SrTiO_3$ = STO. It consists of strontium, titanium and oxygen. While STO has a cubic crystal structure above 105 K [cf. Fig.7(a)], a structural phase transformation towards a

---

[9]Detailed material properties are given in App.B.

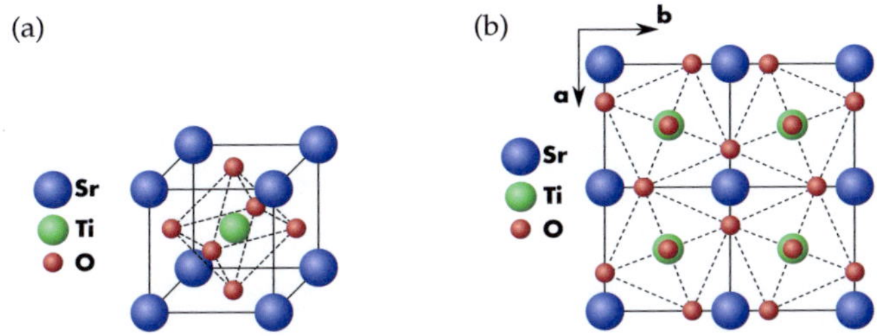

Figure 7: STO crystal structure (a) above 105 K and (b) below 105 K [cf. 107].

tetragonal centrosymmetry takes place below this temperature [cf. Fig.7(b)] [88, 107, 170]. The thermal conductivity of STO is around 10 W/mK, which we consider as moderately-conductive. However, mechanical strain can decrease the thermal conductivity by 50% [164]. Furthermore, a high potential for the Seebeck coefficient S in STO (so that ZT > 2) [125] and a high melting point ($\approx$ 2350 K) [95] make this material a promising thermoelectric.

The second material we focus on is the perovskite material praseodymium calcium manganite $Pr_{1-x}Ca_xMnO_3$ = PCMO. It consists of praseodymium, calcium, manganese and oxygen. A unit cell of this material is shown in Fig.8. Considering the oxygens of the next unit cell, an MnO_6 octaeder can be recognized. This octaeder is tilted and distorted between different unit cells [cf. 48, 53, 54, 85, 155]. In this work, PCMO is investigated for two different applications.

First, PCMO is interesting for thermoelectrics because it exhibits a low thermal conductivity of around 1.5 W/mK and an electron-phonon coupling [85]. Here, the electron-phonon coupling originates in polarons. A polaron is regarded as a charge carrier, releasing lattice vibrations (phonons) by moving through the crystal structure. This electron-phonon coupling turns this material into a desirable thermoelectric to understand phonon movement and interaction.

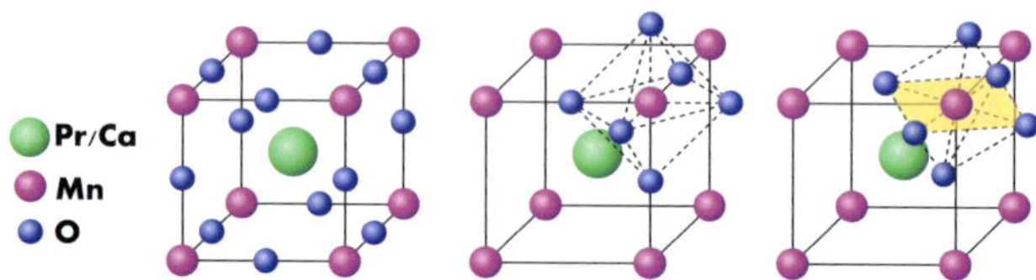

Figure 8: Crystal unit cell of PCMO including the MnO_6 octaeder with its tilt and distortion [cf. 85].

Second, micro- to nanoscale PCMO samples exhibit a change in electric resistance by electric stimulation, such as impulse induced high current densities [cf. 36, 49, 108]. Above a certain current density, the electric resistance changes and remains afterwards the current pulse is switched off. Typically, pulse durations are 100 ns up to 1 s [cf. 59, 104]. The electric resistance obtained in this way, can further increase or decrease, depending on the polarity of the voltage of the next applied pulse and the magnitude of the current [108]. Based on this behaviour, a so-called resistive switching effect occurs [cf. 7, 59, 89, 90, 104, 105, 108]. Due to this reversible resistance change, PCMO is interesting for the application in memory devices [39]. A possible arrangement of multiple resistive switching materials for resistive random access memories (RRAM) is shown by Nauenheim et al. [120]. They pattern sandwich structures of nanoscale resistive switching materials between crossing bars so that an array exists as shown in Fig.9. Although they apply a different resistive switching material, this structure should also be feasible with PCMO.

However, the exact influences on the switching mechanism are still unclear [cf. 49]. For instance oxygen migration in the PCMO material could contribute to the resistive switching [105]. But, this migration process would require a significant temperature increase to initiate migration. However, it is absolutely unclear which

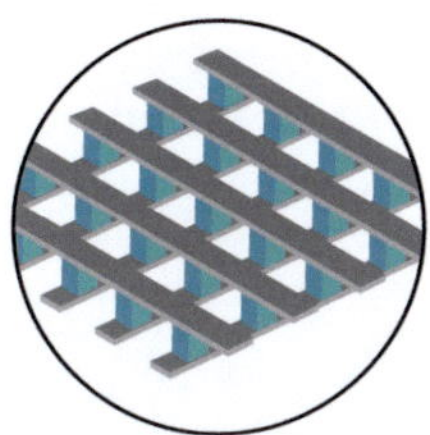

Figure 9: Example for an arrangement of multiple resistive switching materials for resistive random access memories [cf. 120]. The grey bars are the top and bottom contacts for the electric current and the blue blocks represent the resistive switching material.

temperatures occur for such high current densities and whether the occurring temperatures have significant influence on the temperature dependent resistivity or not. Thus here, Joule heating could contribute significantly to a thermally assisted switching [cf. 105]. Since it is exceedingly difficult to measure the temperature inside a small scale PCMO material experimentally, the influence of Joule heating on the electric resistance is studied in this work numerically. In contrast to thin film investigations for thermoelectrics, micro- to nanoscale cylindrical PCMO samples (pillars) are investigated.

1.2 Heat conduction in solids

In general, propagation of heat takes place by three different types of heat transfer, as shown in Fig.10.

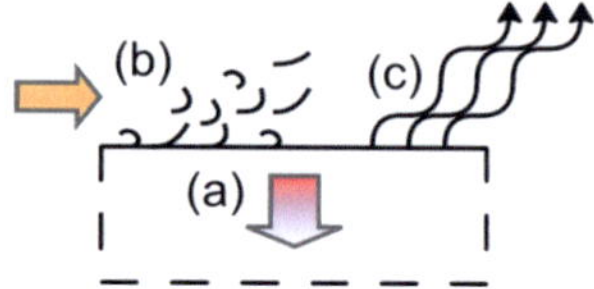

Figure 10: Different types of heat transfer: (a) heat conduction, (b) convection and (c) radiation.

Convection describes heat transfer between two media which are at different temperatures. Here, at least one medium is a flowing fluid. Radiation is heat transfer by electromagnetic waves. In comparison to convection, radiation does not need an additional medium to be involved [21, p. 4]. Thus, radiation can also take place in vacuum. Conduction describes heat transfer through electron and phonon movements and interactions within a body itself [cf. 145, 151]. Conduction takes place when a temperature gradient exists in a material between two different points along an axis as shown in Fig.11.

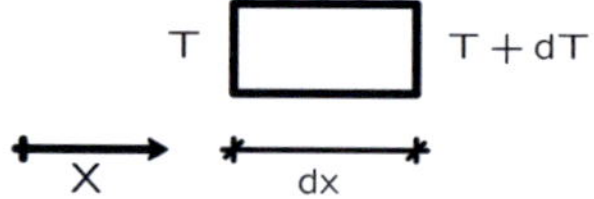

Figure 11: Change of temperature.

According to Carslaw and Jaeger [35, p. 7], the flow of heat, per unit time and area, in the direction of x, is defined as heat flux

$$q_x = -\kappa \frac{\partial T}{\partial x} \, ,$$

(1.5)

where κ is the thermal conductivity of the material and T the temperature. For an isotropic solid, the fluxes in three dimensions are defined as

$$q_x = -\kappa \frac{\partial T}{\partial x}, \quad q_y = -\kappa \frac{\partial T}{\partial y}, \quad q_z = -\kappa \frac{\partial T}{\partial z}. \tag{1.6}$$

Using the gradient operator ∇ and combining the spatial fluxes to the vector identity $\vec{q}$, Fourier's law of heat conduction writes as

$$\vec{q} = -\kappa \nabla T. \tag{1.7}$$

To derive the three-dimensional partial differential equation of heat conduction, an infinitely small cubic volume element is taken [cf. Fig.12].

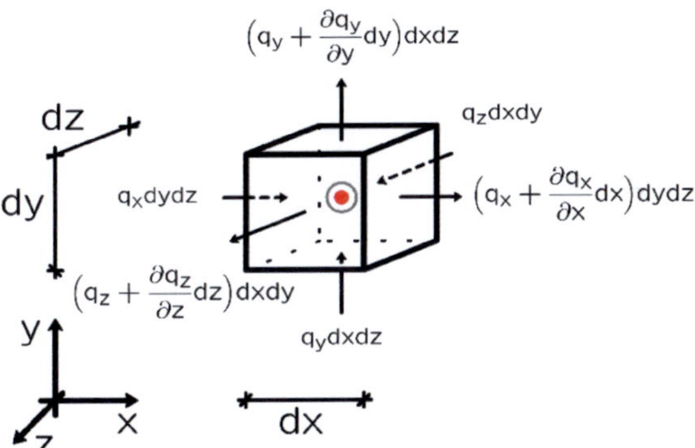

Figure 12: Heat fluxes in three dimensions.

The heat flux balance of Fig.12 results in the stationary circumstance

$$\left(\frac{\partial q_x}{\partial x} + \frac{\partial q_y}{\partial y} + \frac{\partial q_z}{\partial z}\right)dxdydz = 0. \tag{1.8}$$

Time dependent temperature changes are taken into account by the mass density ρ and the specific heat capacity c_p through the term $\rho c_p \frac{\partial T}{\partial t} dxdydz$ [35, p. 9]. Thus, Eq. 1.8 becomes

$$\left(\frac{\partial q_x}{\partial x} + \frac{\partial q_y}{\partial y} + \frac{\partial q_z}{\partial z} \right) + \rho c_p \frac{\partial T}{\partial t} = 0 . \tag{1.9}$$

If heat is generated in the volume, a source term $p(x,y,z,t)$ has to be added to the right hand side [35, p. 10] so that Eq. 1.9 develops to

$$\left(\frac{\partial q_x}{\partial x} + \frac{\partial q_y}{\partial y} + \frac{\partial q_z}{\partial z} \right) + \rho c_p \frac{\partial T}{\partial t} = p(x,y,z,t) . \tag{1.10}$$

Substitution of the heat flux terms by the identities defined in Eq.1.6, the partial differential equation of heat conduction for isotropic solids becomes

$$-\left(\frac{\partial}{\partial x} \kappa \frac{\partial T}{\partial x} + \frac{\partial}{\partial y} \kappa \frac{\partial T}{\partial y} + \frac{\partial}{\partial z} \kappa \frac{\partial T}{\partial z} \right) + \rho c_p \frac{\partial T}{\partial t} = p(x,y,z,t) . \tag{1.11}$$

If κ is constant, rearranging of Eq.1.11 gives

$$\frac{\partial^2 T}{\partial x^2} + \frac{\partial^2 T}{\partial y^2} + \frac{\partial^2 T}{\partial z^2} - \frac{1}{D} \frac{\partial T}{\partial t} + \frac{p(x,y,z,t)}{\kappa} = 0 \tag{1.12}$$

and points out the thermal diffusivity

$$D = \frac{\kappa}{\rho c_p} . \tag{1.13}$$

In case the temperature does not vary with time, Eq.1.11 simplifies to the steady state formulation

$$-\left(\frac{\partial}{\partial x}\kappa\frac{\partial T}{\partial x} + \frac{\partial}{\partial y}\kappa\frac{\partial T}{\partial y} + \frac{\partial}{\partial z}\kappa\frac{\partial T}{\partial z}\right) = p(x,y,z) \qquad (1.14)$$

in which time-dependent parts do not exist. The previously discussed equations are valid for isotropic thermal conductions. However, naturally or artificially created materials can exhibit an anisotropy of thermal conductivity in different spatial directions. The corresponding heat conduction equations are discussed in Sec.3.1.1. Furthermore, it has to be mentioned that in general the thermal conductivity of one material is not constant but varies with temperature [cf. 35]. However, we neglect this effect in this work.

2 The 3ω-method

2.1 Concept, measurement principle and geometry configurations

Concept

The 3ω-method is an alternating current(ac) technique to measure the thermal conductivity of micro- to nanoscale samples. This method finds its origin in the early 20th century. According to literature [133], Corbino [44] discovered a third harmonic component in the voltage signal. Since then, various authors have exploited alternating current techniques to determine thermal properties of samples.[10] For example, Birge et al. [16, 17] used 3ω-detection to measure the specific heat of super cooled liquids, such as glycerol and propylene glycol, near the glass transition phase in the 1980s. Approximately one decade later, Cahill [28, 34] derived the analytic solution of the temperature amplitude for a heater substrate system where the heater has a finite width. Contemporary literature dealing with the 3ω-method refers mostly to Cahill's prominent work.

Measurement principle

The classical 3ω-method [28] uses a thin electrically conducting metal strip deposited on the sample surface to measure the thermal conductivity. The metal strip functions as both the heater and the thermometer. An alternating current

$$I = I_0 \cos(\omega t) \tag{2.1}$$

---

[10]Dames et al. [47] compare different harmonic detections ($1\omega, 2\omega, 3\omega$ signals).

is applied onto the heater. Here, I_0 is the peak current, ω is the angular frequency and t is the time. A classical heater substrate structure is shown in Fig.13. The current is applied at the outer pads of the heater.

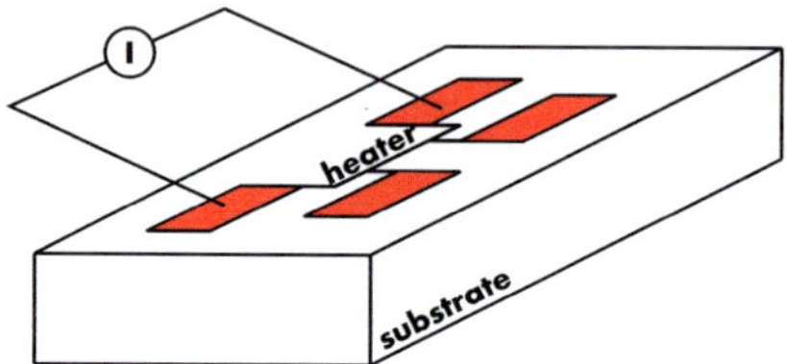

Figure 13: Application of the alternating current at the outer pads.

Passing an alternating electric current through a metal strip with an electric resistance R releases power P as heat, known as Joule heating. The released power writes as

$$P = R \cdot I^2 = R \cdot I_0^2 \cdot \cos^2(\omega t)$$

$$= R \cdot I_0^2 \cdot \frac{1}{2} \cdot (1 + \cos(2\omega t)) \,. \tag{2.2}$$

The released power is shown in Fig.14.

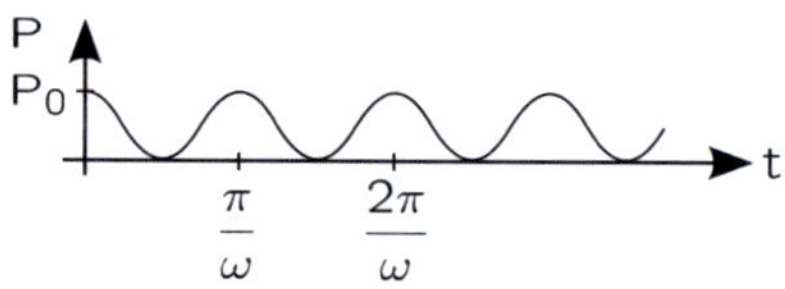

Figure 14: Time dependent released power.

Here, $P_0 = R \cdot I_0^2 \cdot 1/2$ is the applied power amplitude. The time-average mean component of the dissipated power produces a constant temperature gradient with respect to the ambient temperature ($T_0 + T_{co}$), while the alternating component releases diffusive thermal waves which propagate into the sample. Hence, the general temperature rise of the heater and the surrounding material is composed of an average part T_{av} and alternating part $\Delta \check{T}$ due to a constant and an alternating power input, respectively. The general temperature behavior is shown in Fig.15.

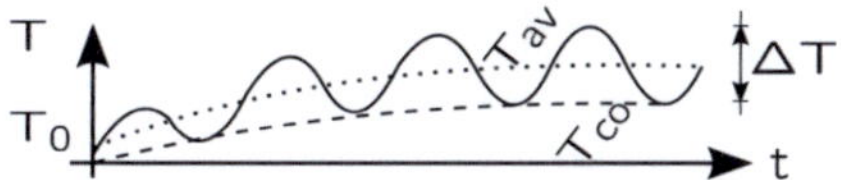

Figure 15: General temperature behavior.

Considering a possible phase shift ς, the temperature rise writes to

$$T = T_{av} + \Delta T \cos(2\omega t + \varsigma). \tag{2.3}$$

Thus, a sufficient time must be given in the measurement until the system reaches a 'swung-in state'. The general temperature profile and the phase shift in the heater depend on the thermal conductivity and thermal mass of the sample material. Due to its temperature dependence, the resistance of the metal strip R oscillates additionally to an average resistance R_{av} at the modulated frequency 2ω with ΔR so that

$$R = R_{av} + \Delta R \cos(2\omega t + \varsigma). \tag{2.4}$$

The inner pads of the metal strip are used to measure a voltage across the heater [Fig.16].
Substitution of Eq.2.1 and Eq.2.4 into Ohm's law leads to a voltage signal U,

$$U = I \cdot R = I_0 R_{av} \cos(\omega t) + \frac{I_0 \Delta R}{2}(\cos(3\omega t + \varsigma) + \cos(\omega t + \varsigma)), \tag{2.5}$$

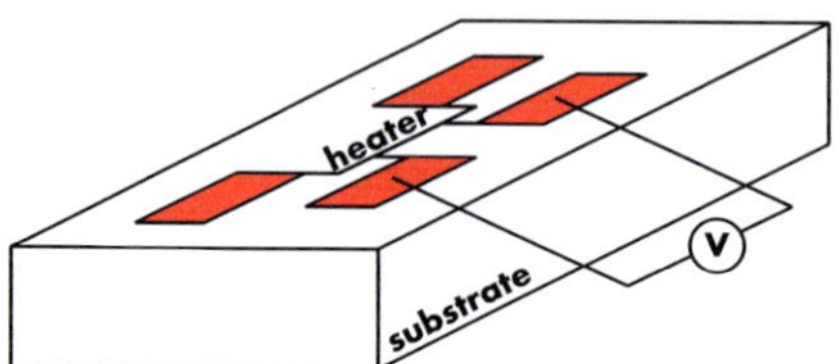

Figure 16: Measurement of the modulated voltage at the inner pads.

containing information at ω and 3ω.[11] Relating the voltage signal to the desired thermal conductivity of the sample, a connection over the temperature amplitude is required. In case of the 3ω-method, it is assumed that the resistance oscillation is linearly connected to the temperature amplitude. This circumstance is shown in Fig.17

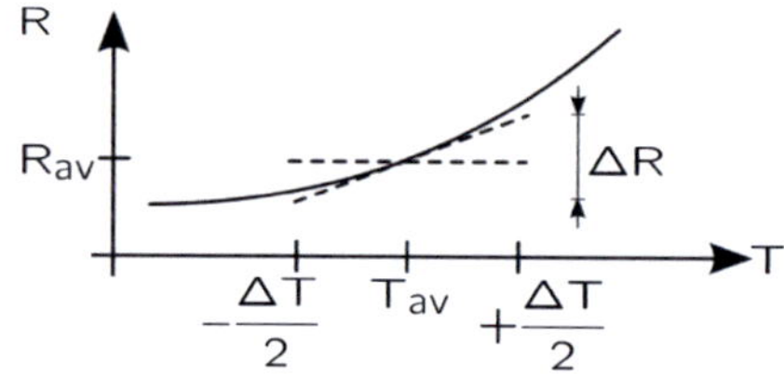

Figure 17: Temperature dependent resistance behavior.

and writes to

$$\Delta R = \frac{dR}{dT} \cdot \Delta T.\qquad(2.6)$$

Here, dT/dR is the specific temperature to resistance behavior of the heater. Rearranging Eq.2.6 and including the third harmonic component of the voltage $U_{3\omega} = I_0\Delta R/2$ and the peak voltage amplitude $U_{1\omega} = I_0 R_{av}$ of the voltage at frequency ω, the temperature

---

[11]Find detailed and further information in App.A.1.

amplitude of the heater is given as [29]

$$\Delta T = \frac{dT}{dR} \cdot \Delta R = 2\frac{dT}{dR} \cdot \frac{U_{3\omega}}{I_0} = 2\frac{dT}{dR}\frac{R_{av}}{U_{1\omega}}U_{3\omega} . \qquad (2.7)$$

This relation connects the voltage measured in experiments to the analytically described temperature amplitude. This connection is necessary to identify the sample's thermal conductivity.

Geometry configurations

The third-harmonic detection is used in a wide range of applications today. Although the working principle is consistent for each approach, different geometry configurations have different analytical descriptions for the temperature amplitude ΔT. One-dimensional like structures, which are mainly free from mechanical supports, such as standing nanopillars [148] or suspended nanowires/microwires [14, 110, 165] are studied or used to determine the thermal properties of surrounding matter such as gas [169] or fluids [87]. Moreover, different groups investigated heater substrate platform structures [cf. Sec.2.3] to determine the thermal conductivity of electrically non-conducting [38, 117] and electrically conducting liquids [40]. In this work, heater substrate platform structures are further investigated for their potential to study bulk-like materials, patterned multilayer structures, and thin films. Also, membrane structures are nowadays used to determine the thermal conductivity of thin films. The group of F. Völklein [159] has been applying membrane structures, made by Micro Electro Mechanical Systems (MEMS-structures), since the 1990s. Various groups are still investigating thermal properties with MEMS where other geometry structures are inappropriate to determine the thermal conductivity of the sample [cf. 58, 79, 82, 146].

Especially the in-plane thermal conductivity is emphasized by membrane MEMS geometries. However, utilizing MEMS, requires either restrictive assumptions, or a three-dimensional analysis for the heat flow. These assumptions concern radiation, heat flow along the heater into the contact pads, and the heat spread in the membrane itself [cf. 77]. To overcome the restrictive assumptions, three-dimensional Finite Element simulations are performed in this work to take into account several aspects of the heat flow. Furthermore, special geometry configurations of the heater structure are presented to determine the cross- and in-plane thermal conductivity.

2.2 Top down geometry

2.2.1 Heat source on bulk materials

The origin for the theoretical analysis of the 3ω-method can be found in the fundamental textbook of Carslaw and Jaeger [35]. They solved the partial differential equation of heat conduction [cf. Eq.1.11] in the field Ω with a point source boundary condition Γ on a semi-infinite half space in two-dimensions [cf. Fig.18(a)]. In the case of the 3ω-method, the applied heat is given as power per length (P_0/L) generated at the frequency $2\omega = 2 \cdot 2 \cdot \pi \cdot f_c$. Here, P_0 is the amplitude of the applied power and considered to be constant. With the assumption that the heater length goes to infinity $L_h(z) \rightarrow \infty$, the substrate is semi-infinite $d_s \rightarrow \infty$, and the heat input is a point source, the temperature oscillation at a distance r from the line source of heat reads as [35, p. 193]

$$\Delta T(r) = \frac{P_0}{L\pi\kappa}K_0^B(qr) \,. \tag{2.8}$$

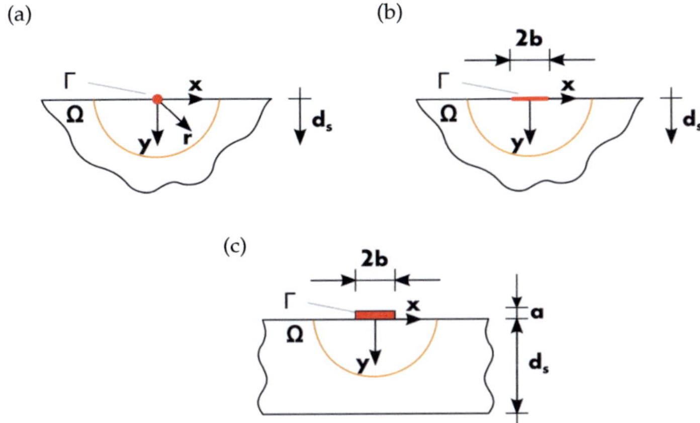

Figure 18: Heat source on bulk materials: (a) Considered system for Carslaw and Jaeger [35]; (b) system for Cahill [28]; (c) system for Borca-Tasciuc et al. [22].

Here, i is the complex number, $r = \left(x^2 + y^2\right)^{\frac{1}{2}}$ is the distance from the line, κ the thermal conductivity of the semi-infinite material, K_0^B the modified Bessel function of second kind and zero order, and $1/q$ is the thermal penetration depth, defined as

$$\frac{1}{q} = \left(\frac{D}{i2\omega}\right)^{\frac{1}{2}}. \tag{2.9}$$

The thermal penetration depth describes the characteristic decay of the temperature amplitude in the material, and thus how deep significant temperature effects propagate into a medium during one cycle of heating [cf. 13, p. 314]. Hence, the ratio between the temperature amplitude at the surface and the temperature amplitude below this depth tends to zero. Consequently, possible changes of material properties below this depth (e.g. different layers of material) have no retroactive influence on the temperature amplitude at the surface. Since the 3ω-method uses a thin metal strip, the heat source has a finite width [cf. Fig.18(b)]. To obtain the temperature amplitude over the heater width, it is convenient [28] to use the Fourier cosine transform of the temperature oscillation Eq.2.8 and the heater width itself. The measurement of the temperature takes place in the heater itself, thus at the surface of the sample. Therefore, y is set to zero so that only the x-coordinate remains in Eq.2.8 [28].

The Fourier cosine transform is defined as [11, p. 3]

$$\mathcal{F}(k) = \sqrt{\frac{2}{\pi}} \int_0^\infty f(x) \cdot \cos(kx)dx\,. \tag{2.10}$$

Based on this definition, the Fourier cosine transform of Eq.2.8 is given as [11, p. 49]

$$\mathcal{F}_p = \sqrt{\frac{2}{\pi}} \cdot \frac{P_0}{L\pi\kappa} \cdot \frac{1}{4} \cdot \frac{1}{\sqrt{(k^2 + q^2)}} \cdot 2\pi = \sqrt{\frac{2}{\pi}} \cdot \frac{P_0}{L2\kappa} \cdot \frac{1}{\sqrt{(k^2 + q^2)}}\,. \tag{2.11}$$

Assuming heat enters the substrate evenly over the heater width, the heater width can be regarded as a rectangular function

$$h = \begin{cases} \frac{1}{b}, & 0 \le x \le b \\ 0, & x > b \end{cases} \tag{2.12}$$

with respect to the symmetry at x = 0, shown in Fig.19 [cf. 27, 111]. Here, b is the half heater width.

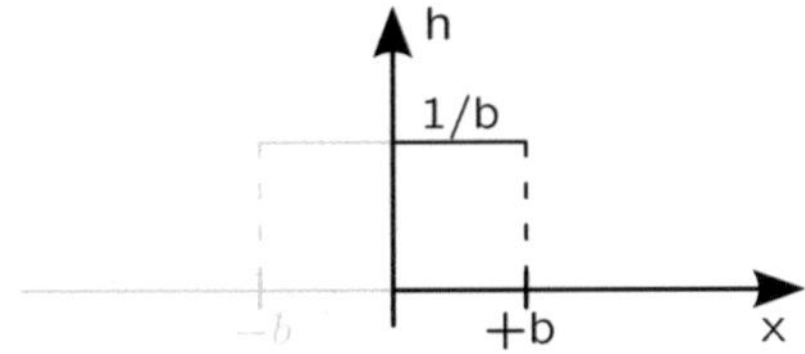

Figure 19: Rectangular function for the heater width.

Using Eq.2.10, the Fourier cosine transform of this rectangular function writes to

$$\mathcal{F}_b(k) = \sqrt{\frac{2}{\pi}} \int_0^b \frac{1}{b} \cdot \cos(kx)dx = \sqrt{\frac{2}{\pi}} \cdot \frac{\sin(kb)}{kb} . \qquad (2.13)$$

In Fourier space, convolution [26, p. 24] of Eq.2.11 and Eq.2.13 results in

$$\mathcal{F}_c(k) = \mathcal{F}_p(k) \cdot \mathcal{F}_b(k) = \frac{2}{\pi} \cdot \frac{P_0}{L2\kappa} \cdot \frac{\sin(kb)}{kb \cdot \sqrt{(k^2 + q^2)}} . \qquad (2.14)$$

The inverse Fourier cosine transform of the convoluted temperature oscillation writes to

$$\Delta T_h(x) = \int_0^\infty \mathcal{F}_c(k) \cdot \cos(kx)dk$$

$$= \frac{P_0}{L\pi\kappa} \cdot \int_0^\infty \frac{\sin(kb) \cdot \cos(kx)}{kb \cdot \sqrt{(k^2 + q^2)}}dk . \qquad (2.15)$$

The latter equation describes the local temperature amplitude from the center of the heater towards its side. However, since the measured temperature amplitude is a mean value over the heater width, it is appropriate to average the temperature amplitude through integration

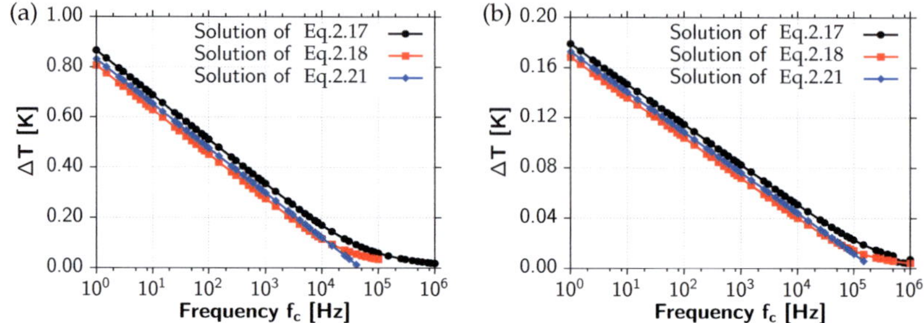

Figure 20: Cahill's solution of Eq.2.17, Eq.2.18 and Eq.2.21 for a heat source on bulk-like materials with $P_0/L = 5$ W/m and $2b = 10$ μm for the substrate materials (a) STO and (b) MgO.

so that

$$\frac{1}{2b} \int_{-b}^{b} \Delta T_h(x) dx = \frac{P_0}{L\pi\kappa} \cdot \frac{1}{2b} \cdot \frac{2}{k} \int_0^{\infty} \frac{\sin(kb) \cdot \sin(kb)}{kb \cdot \sqrt{(k^2 + q^2)}} dk. \qquad (2.16)$$

This is the proposed solution of Cahill [28] for the temperature amplitude measured by the heater

$$\Delta T_h^C = \frac{P_0}{L\pi\kappa_s} \int_0^{\infty} \frac{\sin^2(kb)}{(kb)^2 \sqrt{k^2 + q^2}} dk. \qquad (2.17)$$

Here, κ_s is the thermal conductivity of the substrate. The subscripts h and s indicate heater and substrate respectively. The height and thermal properties of the heater are neglected and no thermal resistance between the heater and the substrate is taken into account.

Moreover, Cahill [28] proposed an approximate solution for the case that the thermal penetration depth is large in comparison to the half heater width ($q^{-1} \gg b \ \rightarrow \ bq \ll 1$).

First, he sets the limits of the integral from 0 to $1/b$ so that

$$\Delta T_h^C = \frac{P_0}{L\pi\kappa_s} \int_0^{1/b} \frac{\sin^2(kb)}{(kb)^2 \sqrt{k^2 + q^2}}\, dk\,. \tag{2.18}$$

Second, he assumes that

$$\frac{\sin^2(kb)}{(kb)^2} = 1\,. \tag{2.19}$$

Underlying these assumptions, Cahill simplifies the general solution of Eq.2.17 to

$$\Delta T_h^C = \frac{P_0}{L\pi\kappa_s} \int_0^{1/b} \frac{1}{\sqrt{k^2 + q^2}}\, dk \tag{2.20}$$

$$= \frac{P_0}{L\pi\kappa_s} \left[-\frac{1}{2}\ln(\omega) - \frac{1}{2}\ln\left(\frac{ib^2}{D}\right) + \text{const} \right]. \tag{2.21}$$

This linearized description contains a frequency dependent real part $\mathrm{Re}(\Delta T_h^C) = [-1/2 \cdot \ln(\omega) + \text{const}]$ and an imaginary part $\mathrm{Im}(\Delta T_h^C) = [-1/2 \cdot \ln(ib^2/D)]$.[12] The frequency dependent temperature amplitude $\mathrm{Re}(\Delta T_h^C)$ is shown in Fig.20 for the solution of Eq.2.17, Eq.2.18 and Eq.2.21.[13] First, the low frequency regime is considered. While the solutions exhibit different axis intercepts, the slope of the ΔT -lines are the same so that these lines are parallel. However, Eq.2.21 shows how the thermal conductivity of an isotropic substrate can be determined from the slope of the real part $\mathrm{Re}(\Delta T_h^C)$ of the complex temperature amplitude versus the logarithm of the frequency.

---

[12] Find App.A.3 for detailed information on the derivation and the constant.
[13] In this work, we only plot the real part of the temperature amplitude $\mathrm{Re}(\Delta T)$ and reduce the labelling of the axis to ΔT.

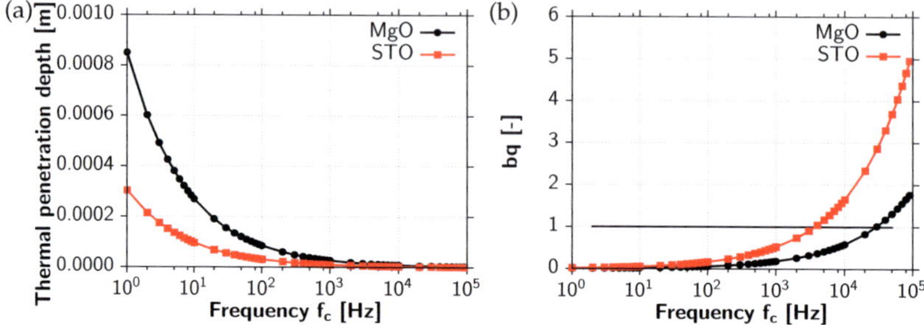

Figure 21: (a) Thermal penetration depth q^{-1} and (b) restriction bq $\ll$ 1 for two different materials.

Given by the linear behavior in the low frequency regime, such as in Fig.:20(a) and Fig.:20(b), the thermal conductivity of the substrate can be extracted by

$$\kappa_s = -\frac{P_0}{L\pi 2} \cdot \frac{1}{\frac{d}{d\ln(\omega)}\mathrm{Re}(\Delta T_h^C)} = -\frac{P_0}{L\pi 2}\frac{d\ln(\omega)}{d\,\mathrm{Re}(\Delta T_h^C)} \tag{2.22}$$

from the slope. Hence, the different axis intercepts do not matter if the thermal conductivity is determined in this way. From a physical perspective, the linearized solution of Eq.2.21 is only valid for frequencies where the temperature amplitudes are greater than zero so that $\Delta T > 0$. Amplitudes equal or smaller than zero can not occur. Second, the high frequency regime is considered. Here, the solutions exhibit a different curve progression. This difference occurs, because at higher frequencies, the thermal penetration depth is on the order of the half heater width [cf. Fig.21(a)]. In this case bq $\ll$ 1 is not fulfilled [cf. Fig.21(b)].

Despite the small thickness of the metal strip, the heat source still has a finite height. Since a heat source with a finite height and width represents a cross-section of a material, it is of interest to take into account its material properties for a more realistic description. Borca-Tasciuc et al. [22] considered this circumstance so that the temperature amplitude over the heater width at the surface of the sample writes to

$$\Delta T_H^B = \frac{\Delta T_h^B}{1 + (\rho_H c_{p_H}) a i 2\omega \Delta T_h^B 2b \frac{L}{P_0}} \, , \qquad (2.23)$$

where

$$\Delta T_h^B = \frac{-P_0}{L\pi\kappa_{crs_1}} \int_0^\infty \frac{1}{\chi} \frac{\sin^2(kb)}{k^2 b^2} \, dk \, . \qquad (2.24)$$

Here, a is the height of the heater, $\rho_H c_{p_H}$ is the thermal mass of the heater and κ_{crs1} is the cross-plane thermal conductivity of the material below the heat source. ΔT_h^B is Borca-Tasciuc's solution of the temperature amplitude without material properties of the heater. The complex term χ contains information about the materials below the metal strip, like the incorporated finite substrate thickness [cf. Fig.18(c)]. For a semi-infinite isotropic substrate this solution is equal to Cahill's solution of Eq.2.17. However, this term is discussed in detail in Sec.2.2.2. It has to be noted, that the thermal conductivity of the metal strip is still neglected. Also a thermal contact resistance between the heater and the investigated material can be regarded in Eq.2.23, but this influence factor is not considered in this work.

A comparison of Cahill's and Borca-Tasciuc's solutions in Fig.22 indicates the differences in the inset for various frequency regimes.

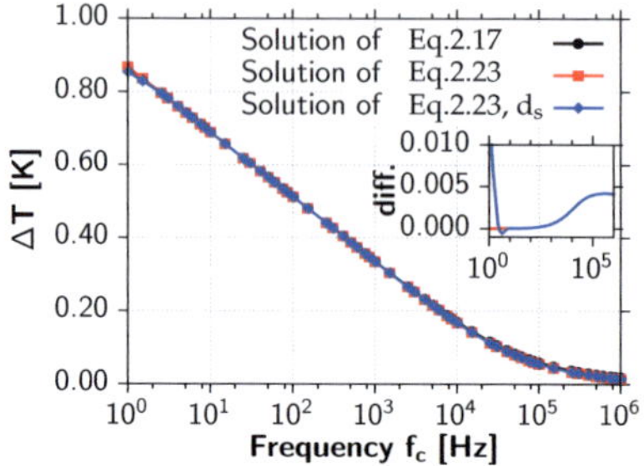

Figure 22: Comparison of Cahill's solution of Eq.2.17 and Borca-Tasciuc's solutions of Eq.2.23 with an infinite and finite substrate. d_s indicates the considered substrate thickness of 1 mm for an STO substrate. The inset contains the detailed difference between the solutions of Eq.2.23 with respect to Eq.2.17. The height of the platinum heater is a = 100 nm.

Eq.2.23 can be solved with respect to an infinite and finite substrate thickness. The solution of Eq.2.23 for a substrate with finite thickness and adiabatic boundary condition differs slightly in the low frequency regime from the solution for an infinite substrate [cf. Fig.22]. Here, the boundary of the substrate interacts with the thermal penetration depth. On the contrary, the thermal penetration depth is small for high frequencies. Thus, the solution for an infinite and finite substrate thickness is coincident in the high frequency regime in the inset of Fig.22. However, in the high frequency regime, the influence of the material properties of the heater become pronounced. This is due to the small thermal penetration depth, being on the order of the heaters width. Traditionally, 3ω-measurements are performed in a frequency regime between 1 Hz and 10 kHz [cf. 28], and the thermal conductivity of the heater and substrate are on the same order of magnitude. Usually, the thermal properties of the heater can be neglected within these boundaries [cf. 152]. However, the limitation of this assumption is discussed more in detail in Sec.2.2.2.

2.2.2 Heat source on layer-substrate materials

Since the 3ω-method uses a thin electrical conducting metal strip deposited on the sample surface, this method is also convenient to determine the thermal conductivity of thin layers which are placed in between the heat source and the substrate material [cf. Fig.23]. Based on steady periodic temperatures at interfaces in composite slaps [cf. 35], Kim et al. [91] derived a formalism to describe monolayer and multilayer systems. Borca-Tasciuc took this formalism and incorporated several options, like the real substrate size, and expanded the solution for anisotropic layered systems, assuming the materials to be infinite in the lateral direction. The complex term χ of Eq.2.24 contains these information. Rewriting Eq.2.24 with $\chi = \varpi_1 \cdot \psi_1$ results in

$$\Delta T_h^B = \frac{-P_0}{L\pi\kappa_{crs_1}} \int_0^\infty \frac{1}{\varpi_1 \cdot \psi_1} \frac{\sin^2(kb)}{k^2b^2} \, dk, \tag{2.25}$$

where the terms ϖ_1 and ψ_1 are:

$$\varpi_{g-1} = \frac{\varpi_g \frac{\kappa_{crs_g} \psi_g}{\kappa_{crs_{g-1}} \psi_{g-1}} - \tanh(\phi_{g-1})}{1 - \varpi_g \frac{\kappa_{crs_g} \psi_g}{\kappa_{crs_{g-1}} \psi_{g-1}} \cdot \tanh(\phi_{g-1})}, \quad g=2...num \tag{2.26}$$

$$\psi_g = \left(\kappa_{in/crs_g} k^2 + \frac{i2\omega}{D_g} \right)^{0.5} \tag{2.27}$$

$$\phi_g = \psi_g d_g \tag{2.28}$$

$$\kappa_{in/crs_g} = \kappa_{in}/\kappa_{crs} \tag{2.29}$$

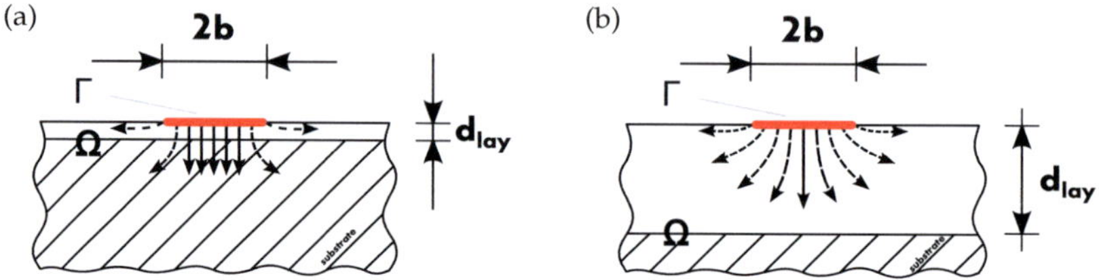

Figure 23: Heat source on layer-substrate materials: Schematic heat spread for (a) broad heaters on thin layers and for (b) narrow heaters on thick layers.

Here, num is the total number of layers (including the substrate) for the corresponding subscript g. The subscript g belongs to the gth layer, starting from top, down to the substrate. The subscript crs indicates the cross-plane value of the corresponding quantity. The material properties of each layer are given by the thickness d_g and the respective thermal diffusivity D_g. The effect of anisotropic thermal conduction is taken into account by $\kappa_{\text{in/crs}}$, the ratio of in-plane to cross-plane thermal conductivity. For a semi-infinite substrate $\omega_{\text{num}} = -1$ while for a finite thickness, $\omega_{\text{num}} = -\tanh(\psi_{\text{num}}d_{\text{num}})$ for an adiabatic and $\omega_{\text{num}} = -1/\tanh(\psi_{\text{num}}d_{\text{num}})$ for an isothermal boundary condition [22].[14] However, the solution of Borca-Tasciuc is rather inconvenient to determine the thermal conductivity of an investigated material, since this convoluted solution is very processor-intensive, especially for anisotropic multilayer structures. But treating the investigated structures with some assumptions for the heat flow, Lee and Cahill [99] proposed an elegant solution to determine the thermal conductivities of thin poorly-conductive dielectric layers.[15]

---

[14] The solutions of Borca-Tasciuc are useful to understand complete system behaviors, but depending on the numbers of layers and other conditions, computation costs can be high.

[15] Also previously, numerous studies were published by Cahill's group, dealing with the determination of thin layer thermal conductivities, such as [32],[30] and [100].

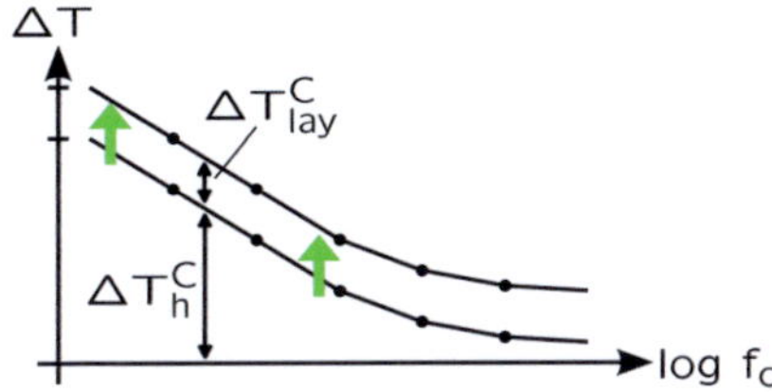

Figure 24: General system behaviour for the offset model.

The so-called offset model [99] regards a poorly-conductive layer as a one-dimensional thermal resistor, so that the cross-plane thermal conductivity can be determined. This model assumes the half heater width to be much greater than the thickness of the layer and the thermal conductivity of the substrate to be much greater than the thermal conductivity of the layer:

$$\left.\begin{aligned} b &\gg d_{lay} \\[4pt] \kappa_s &\gg \kappa_{lay} \end{aligned}\right\} \quad \text{offset model} \tag{2.30}$$

Given these circumstances, a parallel shift along the temperature amplitude axis arises in the low frequency regime onto the solution of a heater substrate system [cf. Fig.24]. Here, the low frequency regime corresponds to the restriction $bq \ll 1$, given in Sec.2.2.1. The overall measured temperature amplitude in the heater ΔT^C_{hls} is now composed of the temperature amplitude by the substrate ΔT^C_h of Eq.2.17 and an additional temperature amplitude of the thin layer ΔT^C_{lay}. This series connection writes to

$$\Delta T^C_{hls} = \Delta T^C_{lay} + \Delta T^C_h . \tag{2.31}$$

With this temperature amplitude offset, the thermal conductivity of the investigated layer can be determined by

$$\Delta T_{\text{lay}}^{\text{C}} = \frac{P_0}{L\kappa_{\text{lay}}} \frac{d_{\text{lay}}}{2b} .$$ (2.32)

Obviously, the temperature amplitude information of the substrate must be given for this procedure. Hence, two measurements are required in general for the investigated material. The first measurement determines the thermal properties of the substrate (and the ΔT over $\log f_c$ plot is recorded). The second measurement is applied on the layer substrate system to obtain the offset and thus determine the thermal conductivity of the layer by Eq.2.32.[16]

Also multilayer and superlattice systems can be characterized with the offset model under the circumstances of Eq.2.30. However, the thickness of each individual layer of the investigated package needs to be small enough in comparison to the heater's width [cf. Sec.4.1.2.2]. The measured parallel offset along the temperature amplitude axis corresponds then to an emergent cross-plane thermal conductivity of the investigated package [cf. 24, 101, 156, 172].

Some material combinations require approaches where the measurement frequencies are beyond the classic linear regime [cf. 126]. Different approaches in the high frequency regime are given by Raudzis [134] (up to 0.2 MHz), Wang et al. [163] (up to 0.5 MHz) and Ahmed et al. [2] (up to 1.0 MHz). Ahmed et al. [2] applied the high frequency range to determine the thermal conductivity of the investigated thin layer with just one measurement. In comparison to Cahill's offset model, here the thermal penetration depth is in the size

---

[16]Several variations of this procedure exist in literature, like the differential method or the determination of the layer's thermal conductivity without a reference sample [3].

of the layer's thickness. A fit is subsequently used to determine the thermal conductivity of the thin layer by the characteristic transition zone in the ΔT plot [cf. Sec.4.1.2]. This approach demonstrates the upper limit of feasible frequencies, but would also require the consideration of the thermal properties of the heater [cf. Sec.4.1.2]. Although the offset model is applied on geometrical systems, where the half heater width is much greater than the thickness of the layer, the slight in-plane heat spread affects the temperature amplitude even for poorly-conductive layers. This in-plane heat spread needs to be taken into account in order to determine accurate thermal conductivities [cf. Fig.23]. Cahill and Lee [33] introduced a correction value in terms of an effective heater width, so that

$$2b_{eff} = 2b + 0.88 \cdot d_{lay} \, . \tag{2.33}$$

This corrected offset model is compared in Sec.4.1.2 with the solution of Borca-Tasciuc and the Finite Element simulations of this work.

In general, a reduction of the heater's width $2b$ towards the dimension of the layer's thickness d_{lay} increases the influence of the in-plane heat spread. Moreover, the influence on the temperature amplitude also increases for moderately- and highly-conductive layer materials. Thus, the in-plane thermal conductivity can be determined in general. Therefore, two measurements are required to distinguish between the cross- and in-plane thermal conductivity. The cross-plane thermal conductivity measurement applies a broad heater while the in-plane measurement applies a narrow heater [cf. Fig.:23].

This approach is used by Borca-Tasciuc et al. [22] to determine the thermal conductivity of 60 μm thick nanochanneled aluminum.

Also, multilayer and superlattice structures with a total thickness of $1-2$ µm are investigated by several groups [cf. 22, 23, 152]. In these references, either a tremendous thickness, or moderately- and highly-conductive layer materials are investigated. But as mentioned in Sec.1, poorly-conductive layer materials are a prerequisite for future thermoelectrics.

However, in this work, the individual thickness of the investigated materials is on the order of $0.1-0.5$ µm and hence up to ten times smaller. Here, the latter approach is no longer suitable, since variations in ΔT are smaller than the measurement accuracies for classic measuring configurations. Given these circumstances, there is still a necessity to develop new measuring configurations for the cross- and in-plane thermal conductivity determination of thin poorly-conductive nanoscale materials.

2.3 Bottom electrode geometry

2.3.1 Heat source between bulk-like materials

In contrast to top down geometries, inverse 3ω-method systems have the heater in between two materials, a so-called bottom electrode geometry. Moon et al. [117] used an inverse application of the 3ω-method to measure the dynamic specific heat and the thermal conductivity of liquids. Based on the one-dimensional heat diffusion, the solution of the temperature amplitude with material on the opposite side of a planar heater is derived under the assumption of semi-infinite materials [cf. Fig.25(a)].

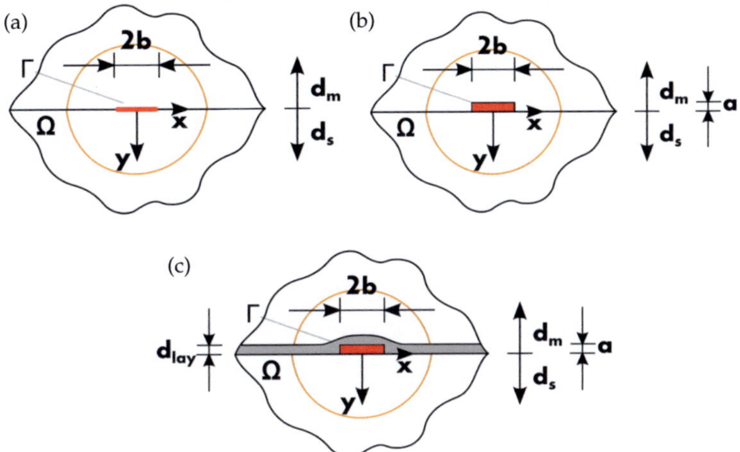

Figure 25: Heat source between bulk-like materials: (a) Considered system for Moon et al. [117]; (b) system for Chen et al. [38]; (c) system for Choi et al. [40].

In this case, the extracted thermal conductivity is a sum of the thermal conductivities of the two materials

$$\kappa_s + \kappa_m = -\frac{P_0}{L\pi 2}\frac{d\ln(\omega)}{d\,\mathrm{Re}(\Delta T_h^C)}\,, \tag{2.34}$$

where, the subscript m indicates the material to be investigated [117]. However, since a real heater has a finite height, this solution is not exact because the ambient material on top encloses the heater [cf. Fig.25(b)]. A so-called boundary mismatch [38] exists due to a heat flow across the side walls of the heater into the liquid. Chen et al. [38], designed a 3ω-method technique to measure the thermal conductivity of liquids and solids, introducing an empirical correction value Θ due to the boundary mismatch to reduce the error for the extracted thermal conductivity.

Thus, Eq.2.34 rewrites to

$$\kappa_s + \Theta \cdot \kappa_m = -\frac{P_0}{L\pi 2}\frac{d\ln(\omega)}{d\,\mathrm{Re}(\Delta T_h^C)}\,. \tag{2.35}$$

Choi et al. [40] proposed a bottom electrode application for the 3ω-method to measure electrically conducting liquids. For this, a thin dielectric layer is deposited on top of the heater to prevent electrical leakage [cf. Fig.25(c)]. Here, the measurements are carried out at low frequencies to ensure the thermal wave penetrates into the liquid.[17] Both, the contour shape of the covering layer and its thermal properties are neglected.

2.3.2 Heater substrate platform

In contrast to top down approaches, where the heat source is always above the investigated material, thin layers on top of the heat source are of special interest in this work. In this thesis, the potential for a new heater substrate platform is investigated for nanostructured materials.[18] By always using the same heater substrate platform as sample holder, an optimum reproducibility of the investigated materials could be achieved. The platform, shown in Fig.26(A), could be utilized to study subsequently patterned multilayer structures (B), covering thin layers (C) or materials placed on top. In this work, placing implies any method to establish thermal contact to the heater such as gluing, pressing, or bonding for materials which cannot be deposited. Examples are brittle crystals (D) or bulk-like materials (E).

---

[17]Similar applications are done for nano- and subnanoliter fluids [124, 127].
[18]Utilizing a heater substrate platform for nanostructured materials was the idea of the joint project and especially the group around Jooss.

In Sec.4.2.2, the sensitivity for the thermal and geometric properties of thin dielectric layers, i.e., changes of the amplitude of the temperature oscillations, in a 3ω-measurement is investigated. For systems where material is placed on top, it is shown how thermal and geometric properties affect the measured thermal conductivity in general and how partial contact of the material affects the temperature amplitude.

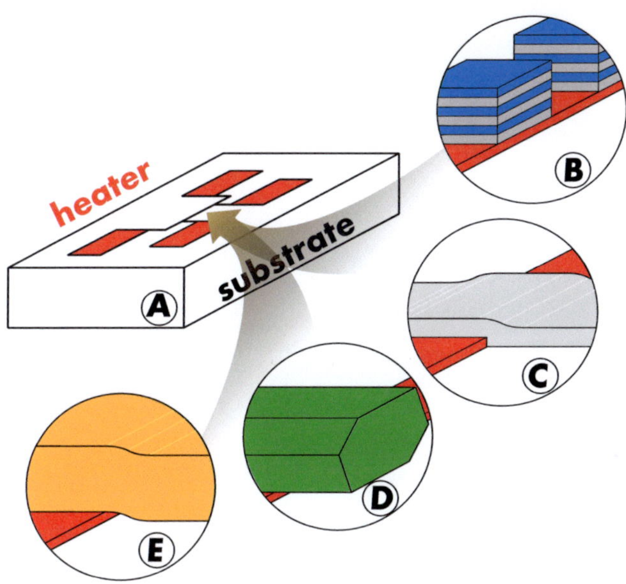

Figure 26: The utilized (A) heater substrate platform for (B) subsequent patterned multilayer structures, (C) a covering thin layer, (D) a placed crystal on top or (E) a pressed on material.

3 Finite Element Model

Finite Element modeling has two decisive advantages for the thermal conductivity determination of nanoscale materials. First, multiphysical coupling of several macroscopic influence factors allows studying derivations from analytical descriptions and hence lead to high precision analysis. Accordingly, the overall understanding of the 3ω-method can be furthered. Second, new measuring configurations can be developed where no analytic solutions are available and thus open new possibilities for nanostructured materials.

In general, the partial differential equation of heat conduction has to be solved. Since the 3ω-measurement takes place in a steady state oscillation level, where the temperature amplitude ΔT oscillates upon a constant temperature gradient [cf. Sec.2.1], one could just perform a harmonic analysis. However, this work applies a combination of steady-state and a time-dependent analysis. By this combination, also the absolute measuring temperatures can be evaluated besides the determination of the temperature amplitudes.

In this work, COMSOL Multiphysics® is used for Finite Element analysis. The interior mathematics interface for partial differential equations is applied to implement the required macroscopic conditions. Therefore, the governing equations are deduced in the following sections to the form of their partial differential equation existence.[19]

---

[19] An introduction for the governing equations of the temperature field is also given by Wang et al. [161].

3.1 Governing equations

3.1.1 Temperature field

We distinguish between three different general areas. The area of the substrate, the heater and the investigated material [Fig.27]. If possible, the system simplified through axis of symmetry.

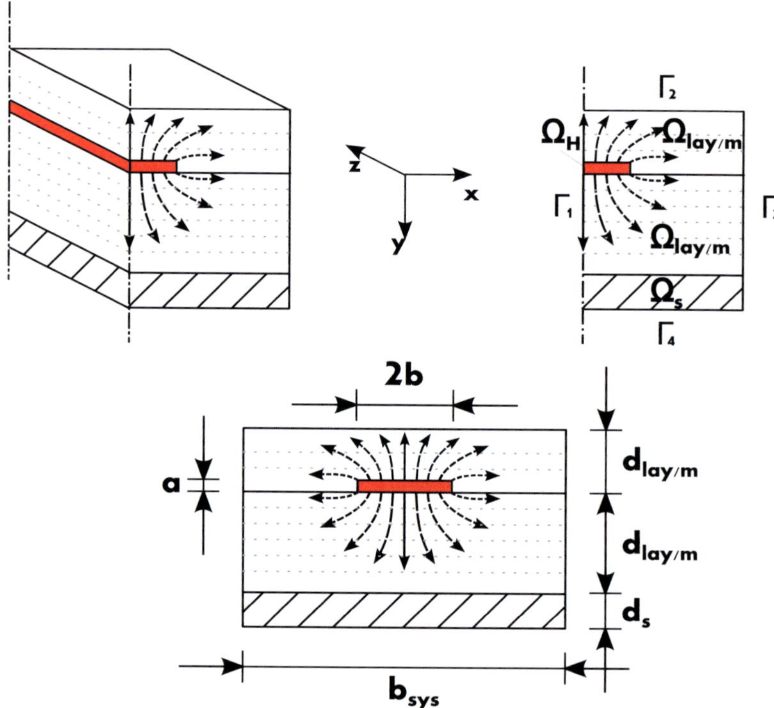

Figure 27: System definition of areas and boundaries.

The substrate is considered to be isotropic and without any source terms for heat dissipation, so that the partial differential equation can be expressed as

$$\Omega_s : \qquad \frac{\partial^2 T_s}{\partial x^2} + \frac{\partial^2 T_s}{\partial y^2} + \frac{\partial^2 T_s}{\partial z^2} - \frac{1}{D_s}\frac{\partial T_s}{\partial t} = 0\,. \tag{3.1}$$

Here D_s is the diffusivity of the substrate. Also the heater is assumed to be isotropic so that

$$\Omega_H : \qquad \frac{\partial^2 T_H}{\partial x^2} + \frac{\partial^2 T_H}{\partial y^2} + \frac{\partial^2 T_H}{\partial z^2} - \frac{1}{D_H}\frac{\partial T_H}{\partial t} + \frac{p}{\kappa_H} = 0\,. \tag{3.2}$$

Here, κ_H is the thermal conductivity of the heater. We take into account a linear temperature coefficient α so that the heat source term p of Eq.:3.2 reads as

$$p = \varrho_0 \cdot (1 + \alpha \cdot (T - T_0)) \cdot j_0^2 \cdot \frac{1}{2} \cdot (1 + \cos(2\omega t))\,. \tag{3.3}$$

Here, T_0 is the ambient temperature and j_0 is the peak current density. Since the 3ω-method applies a power per unit length $W/1m$ in z-direction, but is not specific about the heater's properties, the corresponding current density must be calculated. We calculate the peak current with the specific electric resistivity ϱ_0 at room temperature by

$$\frac{P_0}{L} = \frac{\varrho_0}{A} \cdot I_0^2 \qquad \rightarrow \qquad I_0 = \sqrt{\frac{P_0}{\frac{\varrho_0 \cdot L}{A}}}\,, \tag{3.4}$$

where L is the length of the heater and A is the cross-sectional area of the heater.

The peak current density of Eq.3.3 is now calculated by

$$j_0 = \frac{I_0}{A}.$$

(3.5)

The thermal conductivity coefficients of a three-dimensional anisotropic solid are components of a second-order tensor [cf. 35]. In this work, we distinguish between pure in-plane and cross-plane thermal conduction since we focus on thin layered materials. Hence, the governing equation of the anisotropic investigated material is

$$\Omega_{\mathrm{lay/m}}: \quad \kappa_{m_{\mathrm{in}}}\frac{\partial^2 T_m}{\partial x^2} + \kappa_{m_{\mathrm{crs}}}\frac{\partial^2 T_m}{\partial y^2} + \kappa_{m_{\mathrm{in}}}\frac{\partial^2 T_m}{\partial z^2} - \rho_m c_{p_m}\frac{\partial T_m}{\partial t} = 0.$$

(3.6)

Here $\kappa_{m_{\mathrm{in}}}$ and $\kappa_{m_{\mathrm{crs}}}$ represent the in-plane and cross-plane thermal conductivities respectively. T_m is the temperature and $\rho_m c_{p_m}$ is the thermal mass of the investigated material. In case of two-dimensional simulations, the derivatives in z-direction do not exist so that $\frac{\partial}{\partial z} = 0$. For all areas, the intended measuring temperature represents the initial condition for the temperature

$$\Omega: \quad T(t = 0) = T_{\mathrm{init}},$$

(3.7)

where, T_{init} is the initial temperature and $T(t = 0)$ is the intended measuring temperature at time zero.

Since the investigation of various geometric structures in Sec.4 requires different boundary conditions (BC), these conditions are given in each section respectively.

However, we assume for every investigation ideal thermal contact between all materials [cf. 91] so that

$$T_k = T_{k+1} \, . \tag{3.8}$$

Thus, the interface boundary conditions for the heat flux write to

$$\Gamma_{\text{Int}} : \qquad \kappa_k \frac{\partial T_k}{\partial n} = \kappa_{k+1} \frac{\partial T_{k+1}}{\partial n} \, . \tag{3.9}$$

Here, the subscript k and k+1 indicate two contacting materials. However, it has to be mentioned, that interface effects can influence the heat transfer significantly under special circumstances [31].
The occurring temperature amplitude is determined by

$$\Delta T = T_{\text{mean}_{\text{max}}} - T_{\text{mean}_{\text{min}}} \, . \tag{3.10}$$

Here, $T_{\text{mean}_{\text{max}}}$ and $T_{\text{mean}_{\text{min}}}$ are the maximum and minimum mean temperature within one cycle of heating.[20] The time dependent mean temperature in the heater depends on the spatial dimensions of the investigation. In the following we distinguish between three different cases.
First, for the validation of the Finite Element solution, a heat flux BC is used and thus the time dependent mean temperature is determined over the heater width b by

$$T_{\text{mean}}(t) = \frac{1}{b} \cdot \int_0^b T(t) dx \, . \tag{3.11}$$

---

[20] In the following work, we distinguish between the term mean and average in the manner, that mean refers to a spatial coincident and average to a time dependent coincident.

Second, the time dependent mean temperature for a heater with height a and width b is determined by

$$T_{mean}(t) = \frac{1}{a \cdot b} \cdot \int_0^a \int_0^b T(t) dx dy .$$ (3.12)

Third, in case of a three-dimensional investigation, the time dependent mean temperature is determined by

$$T_{mean}(t) = \frac{1}{a \cdot 2b \cdot L} \cdot \int_0^L \int_0^a \int_0^{2b} T(t) dx dy dz .$$ (3.13)

In contrast to a 3ω-measurement, the PCMO pillar studies exhibit a rotation symmetry since the sample is basically a cylinder. The system and the coordinate system definition is shown in Fig.28.

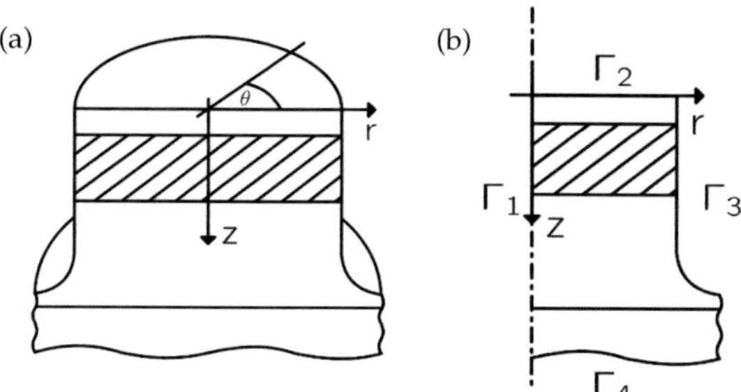

Figure 28: PCMO pillar: 28(a) System and axis definition and 28(b) reduction to cylinder coordinates.

Consequently, the partial differential equations of each material can be transferred to cylindrical coordinates with respect to cosine and sinus functions. Here, we consider every material to be isotropic.

Hence, the respective partial differential equations for the substrate, layer and heating PCMO material become [cf. 35, 132]:

Ω_s :

$$\frac{\partial^2 T_s}{\partial r^2} + \frac{1}{r}\frac{\partial T_s}{\partial r} + \frac{1}{r^2}\frac{\partial^2 T_s}{\partial \theta^2} + \frac{\partial^2 T_s}{\partial z^2} - \frac{1}{D_s}\frac{\partial T_s}{\partial t} = 0 \tag{3.14}$$

$\Omega_{lay/m}$:

$$\frac{\partial^2 T_{lay/m}}{\partial r^2} + \frac{1}{r}\frac{\partial T_{lay/m}}{\partial r} + \frac{1}{r^2}\frac{\partial^2 T_{lay/m}}{\partial \theta^2} + \frac{\partial^2 T_{lay/m}}{\partial z^2} - \frac{1}{D_{lay/m}}\frac{\partial T_{lay/m}}{\partial t} = 0 \tag{3.15}$$

Ω_H :

$$\frac{\partial^2 T_H}{\partial r^2} + \frac{1}{r}\frac{\partial T_H}{\partial r} + \frac{1}{r^2}\frac{\partial^2 T_H}{\partial \theta^2} + \frac{\partial^2 T_H}{\partial z^2} - \frac{1}{D_H}\frac{\partial T_H}{\partial t} + \frac{p}{\kappa_H} = 0 \tag{3.16}$$

In this work, these pillar configurations are studied up to several hundred Kelvin. Here, a linear temperature coefficient is not sufficient to study the temperature development accurately. Therefore, this work applies a non-linear temperature dependent specific electric resistivity which was measured in experiments. The dissipated power is discussed in Sec.4.4 together with the electric resistance behavior. Like for the 3ω-method simulations, we assume here also ideal thermal contact between all materials. The boundary conditions for the outer boundaries are given in Sec.4.4. The mean temperature of the heating PCMO material is determined by

$$T_{mean}(t) = \frac{1}{\pi r_P^2 \cdot d_{PCMO}} \cdot \int_0^{d_{PCMO}} \int_0^{r_P} 2\pi r \cdot T(t) dr dz \,. \tag{3.17}$$

Here, d_{PCMO} is the thickness of the PCMO heating material. The resulting temperature amplitude is again determined by Eq.3.10.

3.1.2 Electromagnetic field

The development of new 3ω-methods includes besides new geometries also temperature amplitude measurements at higher frequencies. However, higher frequencies of the current lead to an increasing tendency of current density distributions towards the surface area in the cross-sectional cut [130]. The reason for this non-uniform distribution is electromagnetic induction, where a time-dependent electric field is accompanied by a magnetic field [130]. The magnetic field decreases from the center of the conductor towards the surface. Accordingly, electrons are pushed towards the surface. Hence, the major part of the electric current flows between the surface and a certain depth level, the so-called skin depth. This occurrence is also known as the skin effect. Such an effect could result in different heating characteristics for a 3ω-measurement in the higher frequency regime [cf. Fig.29].

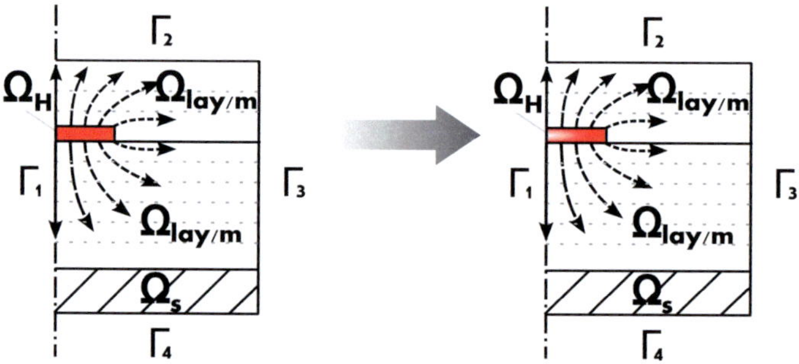

Figure 29: Changing heating distribution due to skin effect.

To study this influence factor, the Maxwell equations [115] are introduced and the electromagnetic wave properties derived and identified in a so-called complex Helmholtz-equation [cf. 61].

The set of Maxwell equations are based on profound discoveries of several scientists in the 18th century and especially the findings of Faraday and Ampere in the 19th century [76]. These equations contain the fundamental information to describe all classic electromagnetic relations. An extensive description for the interaction of electric and magnetic fields is given in [76], [50] and [66], as well as conversions between different unit systems. Although the 3ω-method is performed on a thin metal strip, its dimensions are considerably larger than the atomic length scale and therefore, we consider the Maxwell equations at the macroscopic level. The Maxwell equations in differential form are:[21]

$$\nabla \times \mathbf{H} = \frac{\partial \mathbf{D}}{\partial t} + \mathbf{j} \qquad (3.18) \qquad\qquad \nabla \times \mathbf{E} = -\frac{\partial \mathbf{B}}{\partial t} \qquad (3.20)$$

$$\nabla \cdot \mathbf{D} = \rho_e \qquad (3.19) \qquad\qquad\qquad \nabla \cdot \mathbf{B} = 0 \qquad (3.21)$$

The first equation [Eq.3.18] is known as the extended Ampère's law. Here, $\nabla \times$ represents the rotation operator, $\mathbf{H}$ is the magnetic field intensity, $\mathbf{D}$ is the electric flux density, also known as the dielectric displacement and $\mathbf{j}$ is the electric current density. The bolt notations indicate vector quantities. As a consequence, electric currents and dielectric displacements induce changes in the magnetic field. Without the dielectric displacement term, Eq.3.18 reduces to stationary incidents. In Eq.3.19, $\nabla \cdot$ represents the divergence of a certain quantity and ρ_e is the electric charge density. This equation relates the electric charge to the resulting electric field.

---

[21] For a concise derivation of the partial differential equation, which has to be solved for the electric field, typical operator notation is used.

Eq.3.20 is the Faraday's law of induction. Here, **E** is the electric field intensity and **B** is the magnetic flux density. This equation states that a time dependent magnetic flux density changes the electric field. Eq.3.21 states that the magnetic flux density is solenoidal, so that no magnetic monopoles exist.

The current continuity equation can be obtained by taking the divergence of the extended Ampère's law [Eq.3.18] and subsituting Eq.3.19. Thus, the continuity equation writes to:

$$\nabla \cdot \left(\frac{\partial \mathbf{D}}{\partial t} + \mathbf{j} \right) = \nabla \cdot (\nabla \times \mathbf{H}) \equiv 0 \qquad \rightarrow \qquad \frac{\partial \rho_e}{\partial t} + \nabla \cdot \mathbf{j} = 0 \qquad (3.22)$$

Since a 3ω-measurement takes place at a steady state oscillation level, it is quite convenient to transform the Maxwell equations, under assumptions of time harmonic fields and linear material relations, into their complex form. Assuming that no phase shift exists, the time dependence $\cos(\omega t)$ can be expressed with a complex exponential function[22] to

$$\cos(\omega t) = \frac{1}{2}(e^{i\omega t} + e^{-i\omega t}). \qquad (3.23)$$

It is common to replace the whole cosine just with $e^{i\omega t}$ [130, p. 368]. While the first time derivative of the exponential function $e^{i\omega t}$ is $i\omega e^{i\omega t}$, the time derivatives in the Maxwell equations are just represented by the prefactor $i\omega$. Consequently, time does not appear explicitly. The detailed transformation behavior for this is shown in [130, p. 368].

---

[22] $\cos(z) = \frac{1}{2}(e^{iz} + e^{-iz})$

Hence, the Maxwell equations in complex form are:

$$\nabla \times \mathbb{H} = i\omega\mathbb{D} + \mathbb{J} \qquad (3.24) \qquad\qquad \nabla \times \mathbb{E} = -i\omega\mathbb{B} \qquad (3.26)$$

$$\nabla \cdot \mathbb{D} = \wp_e \qquad\qquad (3.25) \qquad\qquad \nabla \cdot \mathbb{B} = 0 \qquad\qquad (3.27)$$

The blackboard bold typing in Eq.3.18-3.21 indicates the complex quantities. The corresponding current continuity equation writes to

$$-i\omega\wp_e + \nabla \cdot \mathbb{J} = 0. \qquad (3.28)$$

To connect the Maxwell equations to matter, constitutive relations are necessary. The general relations for linear material behaviour write to

$$\mathbb{D} = \epsilon\mathbb{E}, \quad (3.29) \quad \mathbb{B} = \mu\mathbb{H}, \quad (3.30) \quad \mathbb{J} = \sigma_{ec}\mathbb{E}. \quad (3.31)$$

Here, ϵ is the permittivity, μ is the permeability and σ_{ec} is the electrical conductivity. For a linear and isotropic material, the permittivity ϵ and the permeability μ are scalar and constant. Both consist of a vacuum constant and a relative part due to the specific material. Thus, the permittivity is $\epsilon = \epsilon_0\epsilon_r$ and the permeability is $\mu = \mu_0\mu_r$ [66, p. 180, p. 275].

In this work, an approach with the so-called magnetic vector potential $\mathbb{A}_j$ is used to determine the influence of the skin effect [cf. 5, 18–20, 103, 130, 139]. This approach applies the magnetic vector potential as an auxiliary quantity to represent field variables.[23] According to literature [19], the advantage of this potential is to reduce the overall degrees of

---

[23] The derivation and details of this quantity is given in [57, p. 14-1].

freedom and thus save computing costs. Already in the $1980 - 1990^{\text{th}}$, this approach was frequently encountered to solve electromagnetic problems [cf. 19, 103]. The magnetic vector potential $\mathbb{A}_j$ [66, p. 416] is introduced to the magnetic flux density so that

$$\mathbb{B} = \nabla \times \mathbb{A}_j \,. \tag{3.32}$$

With this definition of the magnetic flux density, Eq.3.26 rewrites to

$$\mathbb{E} = -i\omega \mathbb{A}_j \,. \tag{3.33}$$

The definition of the magnetic vector potential arises with a certain freedom for the potential. Here, it is possible to add a gradient of a scalar function to the magnetic vector potential [76, p. 140], such as

$$\mathbb{A}_j \rightarrow \mathbb{A}_j' = \mathbb{A}_j + \nabla f \,, \tag{3.34}$$

without changing the magnetic field density since

$$\nabla \times (\nabla f) = 0 \,. \tag{3.35}$$

Furthermore, a transversal gauge transformation is given by

$$\nabla \cdot \mathbb{A}_j = 0 \,, \tag{3.36}$$

which is the so-called Coulomb gauge transformation. The physical statement of Eq.3.36 is that the magnetic vector potential, and thus the fields, are perpendicular to the direction of propagation [50, p. 174]. Now, the magnetic vector potential complies with the wave equations and thus, the extended Ampère's law and Faraday's law can now be rearranged into a form where the magnetic and electric field intensity are completely decoupled [cf. 76].[24]   However, both forms contain information about the electromagnetic wave and thus the skin depth.[25] We use a form of the extended Ampère's law [Eq.3.24] which can be rewritten to

$$(i\omega\sigma_{ec} - \omega^2\epsilon) \cdot A_j + \nabla \times \mu^{-1}\nabla \times A_j = 0, \tag{3.37}$$

where just the magnetic vector potential is the degree of freedom.[26] Essentially, Eq.3.37 can be rearranged and compared to a so-called complex Helmholtz-equation[27] for waves [cf. 61] in which the complex quantity k is identified by

$$-\nabla \cdot (\mu^{-1}\nabla A_j) + k^2 A_j = 0 \tag{3.38}$$

$$k = \sqrt{i\omega\sigma_{ec} - \omega^2\epsilon}. \tag{3.39}$$

According to literature [130, p. 374], terms of higher order can be neglected for $\sigma_{ec} \gg \omega\epsilon$. Since this work utilizes only platinum heaters and the maximum intended frequencies are below 1 MHz, the term $\omega^2\epsilon$ can be disregarded. To disband the concise writing with the operators above, the partial differential equation which has to be solved for the current density distribution in the area of the heater is presented. Under consideration of several operator rules[28], Eq.3.38 becomes

$$\Omega_H : \qquad -\mu^{-1}\left(\frac{\partial^2 A_j}{\partial x^2} + \frac{\partial^2 A_j}{\partial y^2}\right) + k^2 A_j = 0. \tag{3.40}$$

---

[24]Find App.A.3 for detailed information about the decoupled equations.
[25]Find App.A.3 for information about the skin depth.
[26]Find App.A.4 for the detailed derivation.
[27]Detailed information about the complex Helmholtz-equation can be found in [76].
[28]$\nabla^2 f = \nabla \cdot (\nabla f) = \Delta f$

The initial condition for the magnetic vector potential is

$$\Omega_H : \qquad \mathbb{A}_{j_{init}} = \mathbf{0} . \qquad (3.41)$$

To derive the boundary conditions, it is important to recall Eq.3.18. Just the area of the heater is considered and no interaction between the different materials is taken into account. Thus, the continuity for the magnetic field [cf. 66], simplifies from

$$n \times (\mathbb{H}_{mat_1} - \mathbb{H}_{mat_2}) = \mathbb{J} \qquad to \qquad n \times (\mathbb{H}) = \mathbb{J} , \qquad (3.42)$$

where n is the normal to the cross-sectional cut in direction of the current flow [cf. Fig.30]. In Eq.3.42, $\mathbb{J}$ is the current density passing through the Amperian loop ($\oint_C \mathbb{H} \cdot dl = \int_U \mathbb{J} \cdot dU$) and thus can also be seen as a surface current density, perpendicular to the magnetic field intensity. This kind of approach is also used by other authors like Jakubiuk and Zimny [83] and Gasiorski [61]. The notation for the heater's boundaries are:

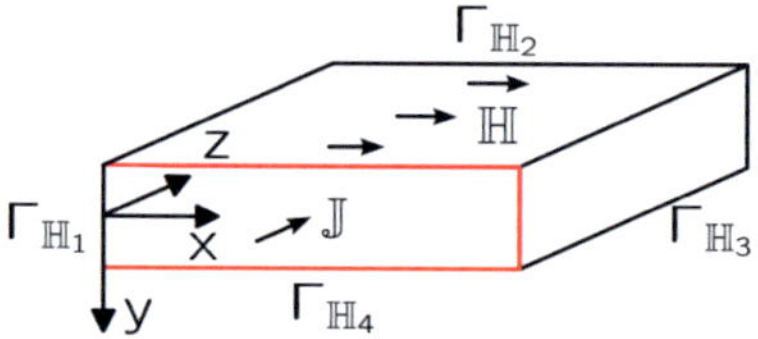

Figure 30: Boundary definitions for the magnetic field intensity and the electric current density.

For the contour of the heater, the equivalent of the peak current is applied [cf. 130] so that

$$\Gamma_{\mathbb{H}_{2-4}} : \qquad \mu^{-1} \left(\frac{\partial \mathbb{A}_j}{\partial x} + \frac{\partial \mathbb{A}_j}{\partial y} \right) = -\frac{I_0}{U} . \qquad (3.43)$$

Here, $U = a + 2 \cdot b$ is the contour of the half heater width. No current is applied for the axis of symmetry, so that

$$\Gamma_{H_1}: \qquad \mu^{-1} \left(\frac{\partial A_j}{\partial x} + \frac{\partial A_j}{\partial y} \right) = 0 \, . \tag{3.44}$$

Since the current density distribution obtained in this way depends on the frequency and the temperature dependent specific electric resistivity, the current density distribution has to be calculated continuously for every time step. The current density in Eq.3.3 is now represented by the effective current density

$$j_{\text{eff}} = \frac{1}{\varrho_0 \cdot (1 + \alpha \cdot (T - T_0))} \cdot (-i\omega A_j) \, . \tag{3.45}$$

3.1.3 Mechanical field

The mechanical system behavior is of interest, especially for thin covering layers, or membrane structures. The eventual exceeding of critical stresses in thin films and in particular the deformation of membrane structures can be examined. This allows besides other determining effects, the investigation of the maximum applicable power per length to limit certain deflections. To derive the partial differential equation, which has to be solved for the mechanical field, a small cut out of a bigger body is considered.[29] For this mechanical subarea equilibrium exists if the volume and surface forces are in balance.

---

[29]Detailed derivation of stress vectors, stress tensor, deformation relations and constitutive relations is given in [67] and [68].

With respect to Cauchy's formula, the condition for mechanical equilibrium [68, p. 11] writes to

$$\nabla \cdot \sigma_{\mathrm{s}} + \mathbf{f} = \mathbf{0}, \tag{3.46}$$

where σ_{s} is the stress tensor, $\mathbf{f}$ represents volume forces and $\nabla \cdot$ is again the divergence operator.[30] Based on the moment equilibrium, the stress tensor σ_{s} is symmetric [68, p. 7]. σ_{s} is a tensor of second order and $\mathbf{f}$ and $\mathbf{0}$ are three-dimensional vectors.

Essentially, thin films must not crack to determine proper thermal conductivity values. Otherwise, the heat conduction could be in an unpredictable path and thus different local thermal properties could influence the measurement. Consequently, just small stresses and deformations are aspired and considered in this work. With this regard and the displacement description of Lagrange, the symmetric infinitesimal strain tensor ε becomes [68, p. 13]

$$\varepsilon = \frac{1}{2}\left(\nabla \mathbf{u} + \nabla \mathbf{u}^{\mathsf{T}}\right), \tag{3.47}$$

where ∇ is the gradient and $\mathbf{u}$ is the mechanical displacement with respect to the spatial direction. Due to symmetry, the strain relations from Eq.3.47 result in:

$$\varepsilon_{xx} = \frac{\partial u_x}{\partial x} = u_{x_x} \qquad \varepsilon_{yy} = \frac{\partial u_y}{\partial y} = u_{y_y} \qquad \varepsilon_{zz} = \frac{\partial u_z}{\partial z} = u_{z_z} \tag{3.48}$$

$$\varepsilon_{xy} = \frac{1}{2}\left(\frac{\partial u_x}{\partial y} + \frac{\partial u_y}{\partial x}\right) \quad \varepsilon_{xz} = \frac{1}{2}\left(\frac{\partial u_x}{\partial z} + \frac{\partial u_z}{\partial x}\right) \quad \varepsilon_{yz} = \frac{1}{2}\left(\frac{\partial u_y}{\partial z} + \frac{\partial u_z}{\partial y}\right) \tag{3.49}$$

---

[30] Forces of inertia are neglected here.

Furthermore, linear elastic material behavior is assumed. The linear stress strain regime is described by Hooke's law. For the one-dimensional case, Hook's law writes to $\sigma_s = E\varepsilon$. A generalization of this law

$$\sigma_s = \mathbf{C} : \varepsilon, \tag{3.50}$$

involves the so-called elasticity tensor $\mathbf{C}$ [68, p. 15],[93]. This tensor represents the elastic properties of the material. In case of isotropic materials, the elasticity tensor is defined by two independent material constants, the so-called Lamé constants. These constants connect the different material properties, like the Young's modulus E_σ, the Poisson's ratio v and the shear modulus $G = E_\sigma/(2 \cdot (1 + v))$.

In general, we assume the 3ω-measurement systems are initially stress free. However, thermal expansions are given fundamentally by the working principal of the 3ω-method. Given a mechanical isotropic material, we assume the thermal strains to be proportional to the temperature differences and equal in all spatial directions. The connection between the strain ε_T and the temperature difference $(T-T_0)$ is given with the linear thermal expansion coefficient to $\varepsilon_T = \beta \cdot (T-T_0)$. Here, ε_T represents a tensor of second order for the thermal expansion coefficient of which just the trace is occupied. Taking into account the thermal expansions, Eq.3.50 can be rewritten in form of the Duhamel-Neumann's law [68, p. 17] to

$$\sigma_s = \mathbf{C} : (\varepsilon - \varepsilon_T) \,. \tag{3.51}$$

Although this work considers only the interaction of temperature leading to expansions, it has to be at least mentioned, that mechanical expansion causes in reality also an entry in the source term of the heat conduction equation [cf. 86, 113].

In general, we assume the sample to be load free for a 3ω-measurement, so that Eq.3.46 changes to

$$\nabla \cdot \sigma_\mathrm{s} = \mathbf{0}. \tag{3.52}$$

Basically this mechanical equilibrium has to be solved with the expanded constitutive relation of Eq.3.51. Depending on the spatial directions of the problem, two or three partial differential equations have to be solved in the time domain. These partial differential equations are for each material [cf. 8]:

$\Omega:$

$$\frac{\partial}{\partial x}\left[2Gu_{x_x} + \frac{2Gv}{(1-2v)}\left(u_{x_x} + u_{y_y} + u_{z_z}\right) - \frac{E_\sigma}{1-2v}\beta(T - T_0)\right]$$
$$+\frac{\partial}{\partial y}\left[G \cdot (u_{x_y} + u_{y_x})\right] + \frac{\partial}{\partial z}\left[G \cdot (u_{x_z} + u_{z_x})\right] = 0 \tag{3.53}$$

$$\frac{\partial}{\partial y}\left[2Gu_{y_y} + \frac{2Gv}{(1-2v)}\left(u_{x_x} + u_{y_y} + u_{z_z}\right) - \frac{E_\sigma}{1-2v}\beta(T - T_0)\right]$$
$$+\frac{\partial}{\partial x}\left[G \cdot (u_{y_x} + u_{x_y})\right] + \frac{\partial}{\partial z}\left[G \cdot (u_{y_z} + u_{z_y})\right] = 0 \tag{3.54}$$

$$\frac{\partial}{\partial z}\left[2Gu_{y_y} + \frac{2Gv}{(1-2v)}\left(u_{x_x} + u_{y_y} + u_{z_z}\right) - \frac{E_\sigma}{1-2v}\beta(T - T_0)\right]$$
$$+\frac{\partial}{\partial x}\left[G \cdot (u_{z_x} + u_{x_z})\right] + \frac{\partial}{\partial y}\left[G \cdot (u_{z_y} + u_{y_z})\right] = 0 \tag{3.55}$$

For all areas, the initial condition represents a displacement free state so that

$$\Omega: \qquad u_x = 0 \quad, u_y = 0 \quad, u_z = 0. \tag{3.56}$$

The corresponding boundary condition for the axis of symmetry is

$$\Gamma_1: \qquad u_x = 0. \tag{3.57}$$

On the contrary, we assume the substrate to be fully clamped to the mount so that

$$\Gamma_4: \qquad u_x = 0 \quad, u_y = 0 \quad, u_z = 0. \tag{3.58}$$

All other outer boundaries are free to move in the spatial directions. Afterwards, the mechanical field is calculated with the given partial differential equations, critical stresses and deformations can be evaluated.

3.2 Transient analysis

In this work, the time dependent process of a 3ω-measurement is treated by transient analysis. A segmentation is chosen for different time regimes, since the temperature amplitude is determined in a steady state oscillation level where the system reached a saturated average temperature. Fig.31 shows the different regimes. In Regime 1, a steady state calculation is applied with a time average mean component of the dissipated power. Thus, we skip the time regime but lift the system from ambient to measuring temperature. Regime 2 proceeds with a transient calculation for a certain number of temperature oscillations with low-time resolution. Thereby, fluctuations of temperature in the whole system are induced to represent a realistic measuring condition. In Regime 3, a transient calculation with a high-time resolution is used to determine accurately the temperature amplitude ΔT.

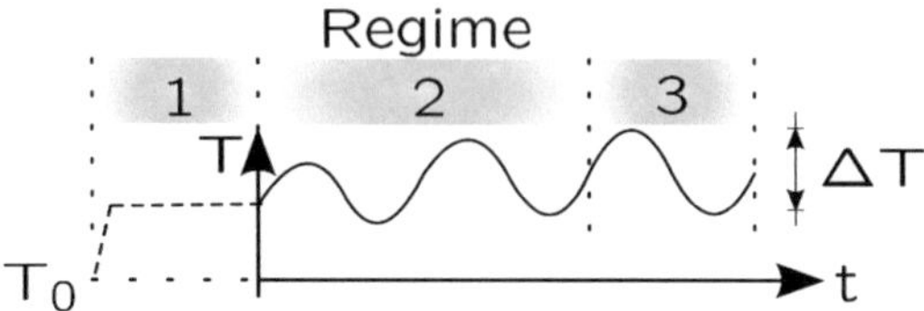

Figure 31: Segmentation of the transient analysis into three different regimes.

Regime 2 and 3 apply the Backward Euler method with strict time stepping for time integration. The Backward Euler method is implemented in COMSOL Multiphysics®. The corresponding step sizes sdp_{2-3}, regime lengths fdp_{2-3} and numerical integration tolerances are defined in Tab.1.

Regime	step size	regime length	tolerance
1			1e-03
2	$sdp_2 = 0.025/f_c$	$fdp_2 = 30.25/f_c$	1e-04
3	$sdp_3 = 0.003/f_c$	$fdp_3 = 30.80/f_c$	1e-05

Table 1: Time and tolerance definitions for the different regimes in a 3ω-measurement simulation.

On the contrary, the analysis of the nanoscale pillars involves only transient calculations. For every simulation a pulse repetition rate of $f_p = 5\,s$ is realized. The course of the applied power is shown in Fig.32(a). The pulse duration varies for certain studies between $dur_p = 1\,\mu s - 1\,s$. In Fig.32(b), the segmentation is given for the temperature development. Here, the corresponding step sizes sdp_{1-12} and regime lengths fdp_{1-12} are defined in Tab.2. A numerical integration tolerance of 1e-12 is set for all pillar studies.

All 3ω-method and pillar simulations apply in every regime the direct solver PARDISO. This solver is implemented in COMSOL Multiphysics®.

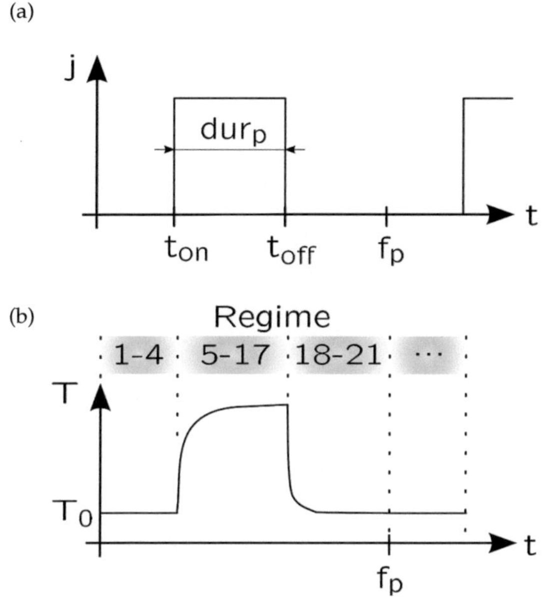

Figure 32: (a) Application of current density and (b) segmentation of the different regimes for the temperature development.

3.3 Eigenfrequency analysis

The 3ω-method is an alternating current technique that applies a time dependent harmonic current $I(\omega t)$ with the angular frequency ω. Based on the alternating current, power is released with two times the current's angular frequency $P(2\omega t)$. Since the mechanical displacement of a material depends on thermal expansion, every 3ω-measurement induces mechanical vibrations with twice the angular frequency of the current. If a system is able to oscillate in motion, it has at least one natural frequency. This particular frequency is called the eigenfrequency.

Regime	step size	regime length
1	$sdp_1 = 1e\text{-}03$	$fdp_1 = 3e\text{-}03$
2	$sdp_2 = 1.4e\text{-}06$	$fdp_2 = 3.0014e\text{-}03$
3	$sdp_3 = 2e\text{-}08$	$fdp_3 = 3.00148e\text{-}03$
4	$sdp_4 = 1e\text{-}09$	$fdp_4 = 3.0015e\text{-}03$
5	$sdp_5 = 1e\text{-}13$	$fdp_5 = 3.001500001e\text{-}03$
6	$sdp_6 = 1e\text{-}12$	$fdp_6 = 3.00150001e\text{-}03$
7	$sdp_7 = 1e\text{-}11$	$fdp_7 = 3.0015001e\text{-}03$
8	$sdp_8 = 1e\text{-}10$	$fdp_8 = 3.001501e\text{-}03$
9	$sdp_9 = 1e\text{-}09$	$fdp_9 = 3.00151e\text{-}03$
10	$sdp_{10} = 1e\text{-}08$	$fdp_{10} = 3.0018e\text{-}03$
11	$sdp_{11} = 1e\text{-}07$	$fdp_{11} = 3.003e\text{-}03$
12	$sdp_{12} = 1e\text{-}06$	$fdp_{12} = 3.01e\text{-}03$
13	$sdp_{13} = 1e\text{-}05$	$fdp_{13} = 3.5e\text{-}03$
14	$sdp_{14} = 1e\text{-}04$	$fdp_{14} = 5.0e\text{-}03$
15	$sdp_{15} = 1e\text{-}03$	$fdp_{15} = 1.0e\text{-}02$
16	$sdp_{16} = 1e\text{-}02$	$fdp_{16} = 1.0e\text{-}01$
17	$sdp_{17} = 1e\text{-}01$	$fdp_{17} = 1.0e\text{+}00$
18	$sdp_{18} = 1e\text{-}02$	$fdp_{18} = 1.1e\text{+}00$
19	$sdp_{19} = 1e\text{-}01$	$fdp_{19} = 1.5e\text{+}00$
20	$sdp_{20} = 5e\text{-}01$	$fdp_{20} = 2.0e\text{+}00$
21	$sdp_{21} = 1e\text{+}00$	$fdp_{21} = 5.0e\text{+}00$

Table 2: Time definitions for the different regimes at nano/micro pillar studies.

The corresponding shape of motion is called the eigenmode [cf. 60, p. 124]. However, if the excitation frequency equals the eigenfrequency of the used samples, a resonance catastrophe could occur. In resonance, a system exhibits especially for the case of small damping large amplitudes in vibration. As result, mechanical displacements with large amplitude could occur which damage the small scale samples. To study the eigenfrequencies, the partial differential equation of motion of structural mechanics has to be solved.

In this work, damping is neglected so that

$$\nabla \cdot \sigma_s + \mathbf{f} = \rho \ddot{\mathbf{u}}, \qquad (3.59)$$

has to be solved [68, p. 11]. Here, double overdots denote the second derivative with respect to time. Moreover, we assume the force vector $\mathbf{f}$ to be zero to solve the problem of natural frequencies [cf. 128]. To obtain the eigenfrequencies, we seek a solution in the harmonic form of

$$\mathbf{u} = \mathbf{\Upsilon} \cdot \sin(\omega t), \qquad (3.60)$$

where $\mathbf{\Upsilon}$ is the amplitude of motion [128, p. 12].

It has to be mentioned, that the geometry and the boundary conditions can have significant influence on the obtained eigenfrequencies. The respective boundary conditions and geometric dimensions are given for each studied system in the respective section.

3.4 Mesh-block building system

Finite Element simulations always involve the challenge to find the balance between accuracy and degrees of freedom. This concerns besides careful considerations about the transient analysis also the Finite Element mesh. It is necessary to adopt the element size, element shape and shape-functions for certain macroscopic and geometrical problems. In this work, quad elements with linear shape functions are used. However, in order to achieve high precision analysis, this work applies special geometry configurations to obtain a pre-defined mesh.[31] Upon the geometry lines a mapped meshing method is applied.

---

[31] The special geometry configurations are generated with Altair HyperWorks®.

The essential advantage of the pre-defined mesh is a highly comparable temperature amplitude in the heater or the pillar, because the mesh constructed in this way, has almost no influence on the temperature amplitude and hence a high comparability and reproducibility is provided. Accordingly, even small deviations by macroscopic influence factors onto the heater can be compared. Fig.33 shows the geometry lines and illustrates the created mesh-blocks.

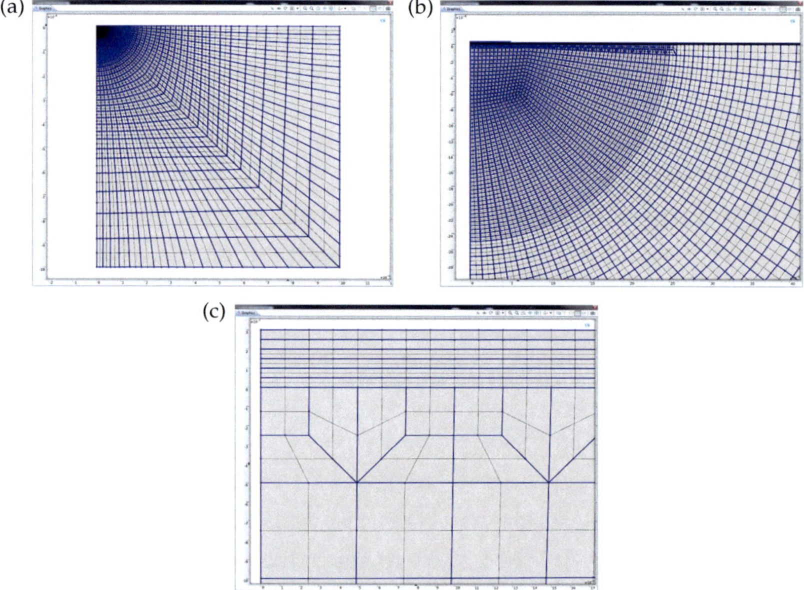

Figure 33: Mesh-block system with geometry lines (blue) and applied mesh (gray). Fig.(b) and (c) zoom in the upper left part of Fig.(a).

4 Investigation of varying structures and multiphysical couplings

4.1 Top down geometry

4.1.1 Heat source on bulk materials

This section examines several influence factors on a 3ω-measurement for heat sources on bulk materials. First, a validation of the used Finite Element Model is given for a quasi semi-infinite half space [cf. Fig.34(a)]. Afterwards, the substrate is reduced to realistic dimensions and various heat transfer conditions for the mount are studied [cf. Fig.34(b)]. Subsequently, we investigate step by step the spatial extent of the heater and take into account its material properties [cf. Fig.34(c)], the temperature dependent resistivity, three-dimensional heat conduction in the substrate, radiation, occurring surface roughness, the skin effect, and thermal expansions and stresses. The following sections deal with the oxide substrate materials MgO, STO and YSZ.[32]

4.1.1.1 Validation of Finite Element Model

First of all, the solution of the Finite Element Model for the temperature amplitude has to be validated by the fundamental solution of Eq.2.17 from Cahill. To check the Finite Element Model by the semi-infinite half space solution of Cahill, a quasi semi-infinite half space with a tremendous size of $d_s = 0.5$ m is modeled.

---

[32] Find App.B for further information.

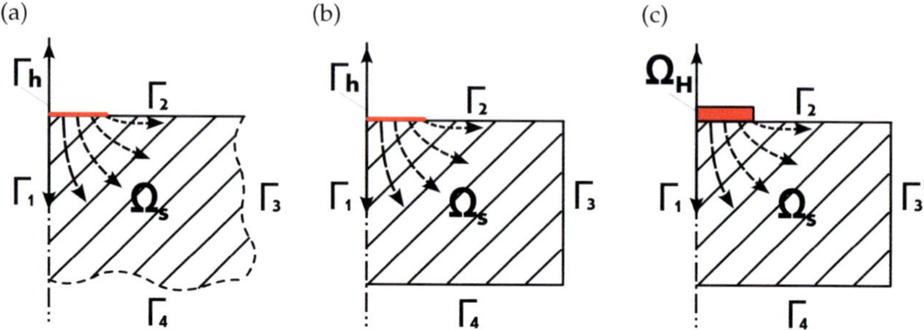

Figure 34: Considered systems for the Finite Element simulations: (a) Quasi semi-infinite half space; (b) real substrate size with heat transfer coefficient; (c) spatial heater with material properties.

The boundaries [cf. Fig.34(a)] Γ_{1-3} are adiabatic

$$\Gamma_{1-3}: \qquad \kappa_{s/h}\frac{\partial T_{s/h}}{\partial n} = 0, \tag{4.1}$$

while Γ_4 is isothermal

$$\Gamma_4: \qquad T_0 = 300\ \text{K}. \tag{4.2}$$

Here the released power per length is applied as heat flux boundary condition (heat flux BC) at Γ_h for $0 \leq x \leq b$ so that

$$\Gamma_h: \qquad p = \frac{P_0}{2b} \cdot \frac{1}{2} \cdot (1 + \cos(2\omega t)). \tag{4.3}$$

The solution of Eq.2.17 and the Finite Element solution are compared in Fig.35(a). Evidently, the solutions are almost congruent in the low frequency regime. For increasing frequency, there is a slight tendency for the Finite Element Model to obtain a greater temperature amplitude towards the high frequency regime.

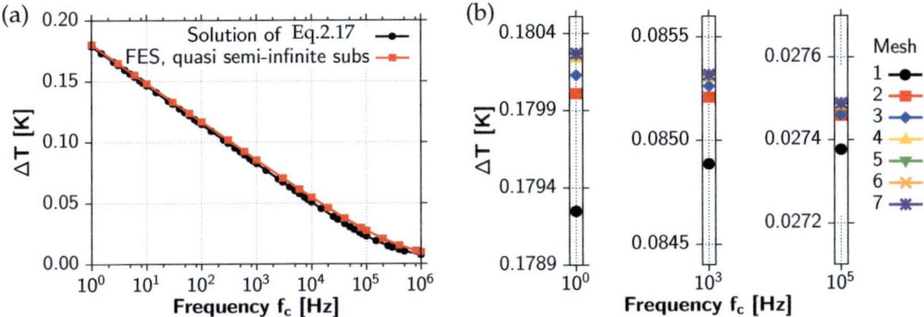

Figure 35: (a) Validation of Finite Element solution (FES) and (b) a corresponding mesh study. The degrees of freedom for Mesh 1 − 7 are 1281, 2923, 4106, 13454, 28581, 49577 and 76310 respectively. The validation is calculated with Mesh 4. The system properties are: $P_0/L = 5$ W/m, $2b = 10$ μm and MgO substrate.

Fig.35(b) shows a mesh study for several applied mesh-block systems. Mesh 4 is chosen as a reasonable compromise between the accuracy and the degrees of freedom because then, the Finite Element solution yields an accuracy on the 4th decimal place. To guarantee this accuracy, an integration tolerance of 1e-05 is realized in COMSOL Multiphysics®. Consequently, it is now possible to compare temperature amplitudes with this numerical precision.

4.1.1.2 Real substrate size

Since a real substrate has a finite thickness [cf. Fig.34(b)], the heat flow from the substrate to the mount needs to be considered. Here, this heat transfer is taken into account with a heat transfer coefficient (htc) so that

$$\Gamma_4: \qquad \kappa_s \frac{\partial T_s}{\partial n} - \mathrm{htc} \cdot (T_s - T_0) = 0. \tag{4.4}$$

The htc is defined as

$$\text{htc} = \frac{q_n/S}{T_s - T_0}\,, \tag{4.5}$$

where q_n/S indicates the flux per surface area. The boundaries Γ_{1-3} are adiabatic. First, an moderately-conductive substrate material is considered to study the influence of thermal contact to the mount at boundary Γ_4. Fig.36(a) shows that a varying heat transfer coefficient can have a significant influence on the temperature amplitude. For instance, even though the thermal penetration depth in STO is three times smaller than the substrate's thickness, there is obviously an influence on ΔT at $f_c = 1\,\text{Hz}$ with a corresponding $q^{-1} = 0.0003\,\text{m}$ [cf. Fig.21(a)]. However, the measured temperature amplitude is less affected for higher frequencies.[33] This circumstance is given by the fact, that the temperature amplitude does not vary for increasing htc above frequencies of $f_c = 10\,\text{Hz}$. **Thus, we propose to perform measurements at frequencies above $f_c = 10\,\text{Hz}$.[34]**
Furthermore, in Fig.36(a) and 36(b), the mean average temperatures over the heater's width are indicated for several htc for two different substrates. For small htc values, thus poor thermal contact, the mean average temperatures are tremendously higher than for large htc values. **Hence, sufficiently good thermal contact must be given to the substrate mount to avoid high system temperatures.**
Given the htc and frequency restriction, the solution of the temperature amplitude for a quasi semi-infinite substrate and the real substrate size are almost coincident [cf. Fig.37].

---

[33] This result is consistent with Borca-Tasciuc's solution for finite substrate thickness [cf. Sec.2.2.1].

[34] Also Zong et al. [173] propose to measure at least from 10 Hz but with respect to the heat spread in the heater's length dimension [cf. Sec.4.1.1.6].

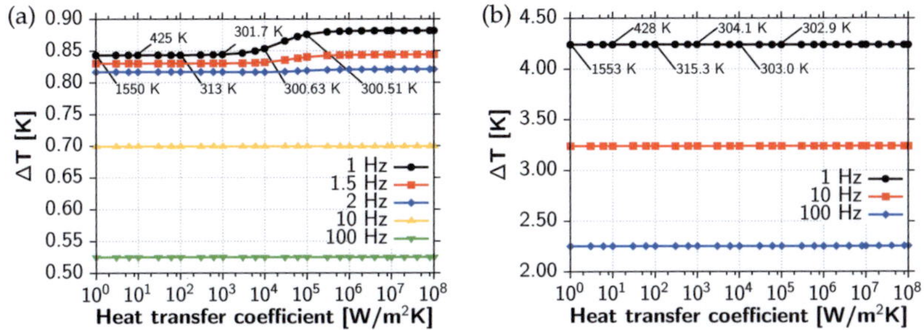

Figure 36: Influence of thermal contact to the mount for a (a) STO and (b) YSZ substrate material. The dimension of the substrate is d_s = 1 mm in height and 2 mm in width. Indicated are in addition the mean average temperatures over the heat source width in time. The system properties are: P_0/L = 5 W/m, 2b = 10 μm and the ambient temperature is 300 K.

4.1.1.3 Geometrical variations of the heater and its material properties

Like the substrate, a real heat source has a finite thickness [cf. Fiq.34(c)]. Typically, a platinum (Pt) heater with thickness arround a = 100 nm is applied. Including the spatial dimensions is of course accompanied by considering its thermal properties. Instead of a boundary condition, the heat is now released through the cross-sectional cut of the heater's area Ω_H by Eq.3.3. The heat input for the heat flux BC model is evenly distributed over the heater width. However, actually the incoming heat can rearrange over the heater's width and height. This circumstance is taken into account in this section. Here, we set α in Eq.3.3 to zero, to develop step by step the different macroscopic influence factors.

In Fig.38(a) and 38(b), a comparison between the BC model and the more realistic heater is given for a highly-conductive and poorly-conductive substrate material respectively. While the ΔT -lines are for a highly-conductive substrate material almost coincident, the

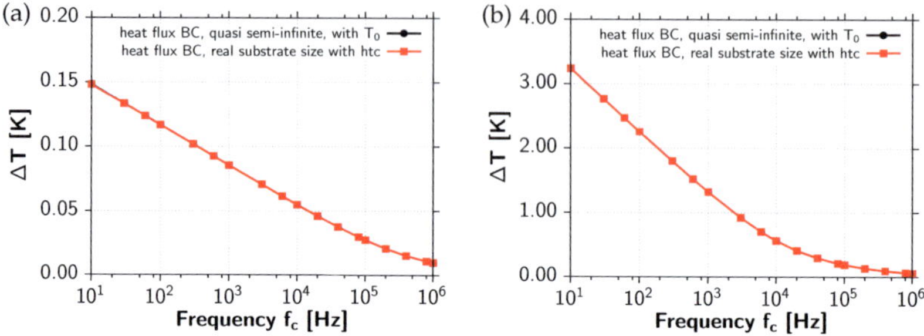

Figure 37: System behavior for a (a) MgO and a (b) YSZ substrate, taking into account a real substrate size. The system properties are: $P_0/L = 5$ W/m, $2b = 10$ µm, $d_s = 1$ mm and htc $= 5 \cdot 10^7$ W/m^{2}K.

lines are slightly different for a poorly-conductive material. This is based on two facts. First, the occurring transversal heat spread in the heater gets more pronounced the bigger the ratio in thermal conductivities of heater to substrate is, which leads to slightly smaller temperature amplitudes [cf. 78]. Second, and most essential, both thermal conductivities (κ of platinum and the substrate) are represented in the slope of the ΔT -line. Thus, one must obtain a slight difference in ΔT even for small amounts of heater material. However, the solution of Cahill in Eq.2.17 and Borca-Tasciuc in Eq.2.23 do not consider the thermal conductivity of the heating material.

Depending on the extension of spatial directions, different affects on the temperature amplitude can be observed. As shown in Fig.39(a), a height variation results almost in the same slope and axis intercept in the low frequency regime. But towards higher frequencies, profound differences can occur [cf. 69, 161]. Based on the small derivations of the different ΔT -lines in the low frequency regime for the same heater width, the thermal properties of the heater are often neglected under the conditions of a classic 3ω-measurement. Because here, the thermal conductivity is determined by the slope of the ΔT -line.

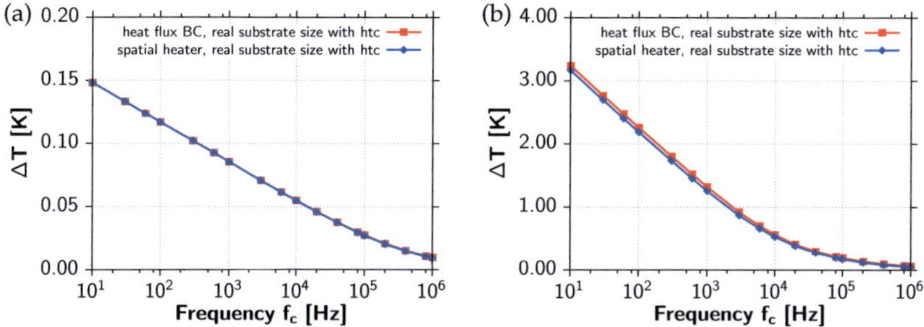

Figure 38: Influence of spatial heater and its material properties for a (a) MgO and (b) YSZ substrate. The system properties are: $P/L = 5$ W/m, $2b = 10$ µm, $a = 100$ nm, $d_s = 1$ mm and htc $= 5 \cdot 10^7$ W/m^{2}K.

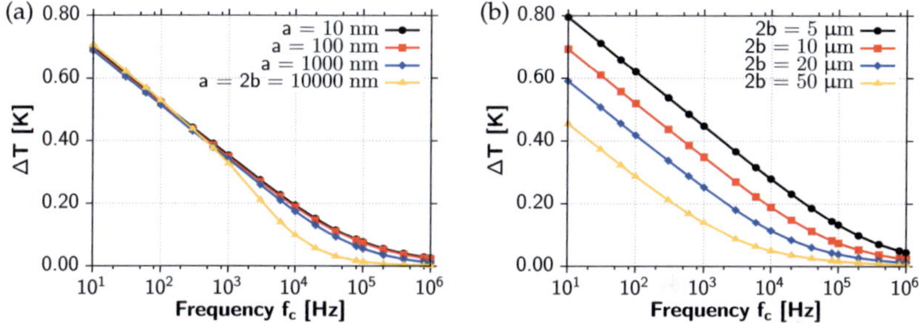

Figure 39: Influence of the heater (a) height and (b) width on the temperature amplitude. The system properties are: $P/L = 5$ W/m, Pt heater, $2b = 10$ µm, $a = 100$ nm, STO substrate, $d_s = 1$ mm and htc $= 5 \cdot 10^7$ W/m^{2}K.

A classic 3ω-measurement applies frequencies up to $f_c = 1000$ Hz, moderately- to highly-conductive substrate materials and a small heater thickness.

On the contrary, a width variation results always in the same slope but different axis intercepts in the low frequency regime [cf. Fig.39(b)].

However, evidentially, high-precision analysis beyond the classic measuring configurations requires to take the heater properties into account. This argument is even further supported in Sec.4.1.2.

4.1.1.4 General temperature distribution for a heater substrate system

In this section, the temperature distribution is examined in general. Therefore, time points are defined within one temperature cycle in the steady state oscillation level in Fig.40.

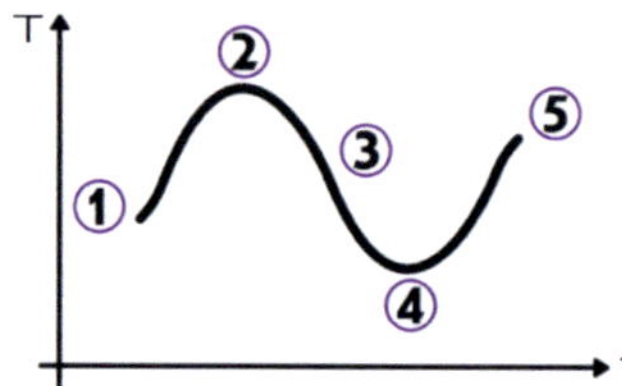

Figure 40: Various time points within one temperature cycle.

First, the maximum and minimum temperatures within a temperature amplitude are considered. The boundary conditions are the same than in the previous section and the source term is given by Eq.3.3.

Fig.41 shows the current temperature at the time points ② and ④. Here, the maximum and minimum temperatures at $f_c = 10$ Hz are in each case higher and lower than for $f_c = 1$ MHz. The lowest mean temperature remains at a higher level for $f_c = 1$ MHz, while the system cools down almost to ambient temperature for $f_c = 10$ Hz. Drawing a horizontal cut between the heater and the substrate along the heater's width and a vertical cut along the axis of symmetry, Fig.42 shows the actual temperatures for the time points ① – ⑤. The horizontal cut in Fig.42(a) and 42(b) exhibits that the maximum temperature is always below the center of the heater. This does not depend on the current's frequency. It illustrates furthermore, the higher minimumn mean average temperature at $f_c = 1$ MHz. While the vertical cut in Fig.42(c) shows clearly decreasing temperature trends for each time point without an overlap, the temperature lines in Fig.42(d) cross each

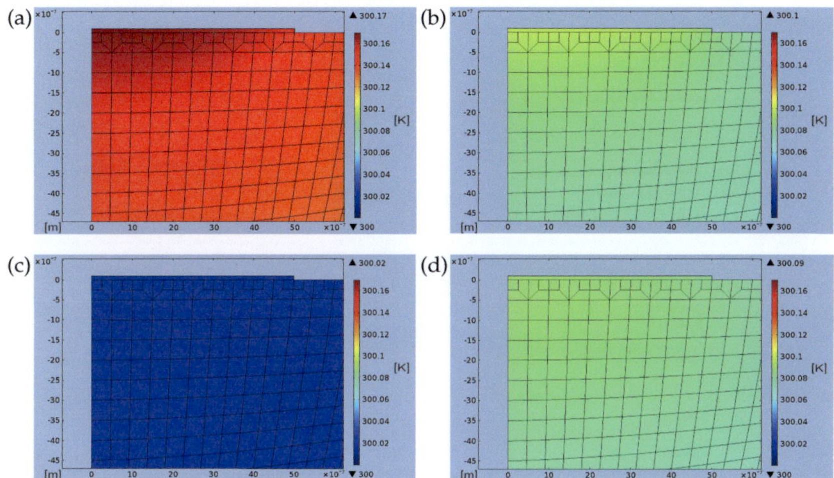

Figure 41: Visualization of (a)(b) the maximum and (c)(d) minimum temperatures within a temperature cycle for $f_c = 10$ Hz on the left and $f_c = 1$ MHz on the right side. The system properties are: $P_0/L = 5$ W/m, Pt heater, $2b = 10$ µm, $a = 100$ nm, MgO substrate, $d_s = 1$ mm and htc $= 5 \cdot 10^7$ W/m^{2}K.

other at a certain vertical depth. The crossing takes place at the thermal penetration depth q^{-1}, indicated by the black vertical line.

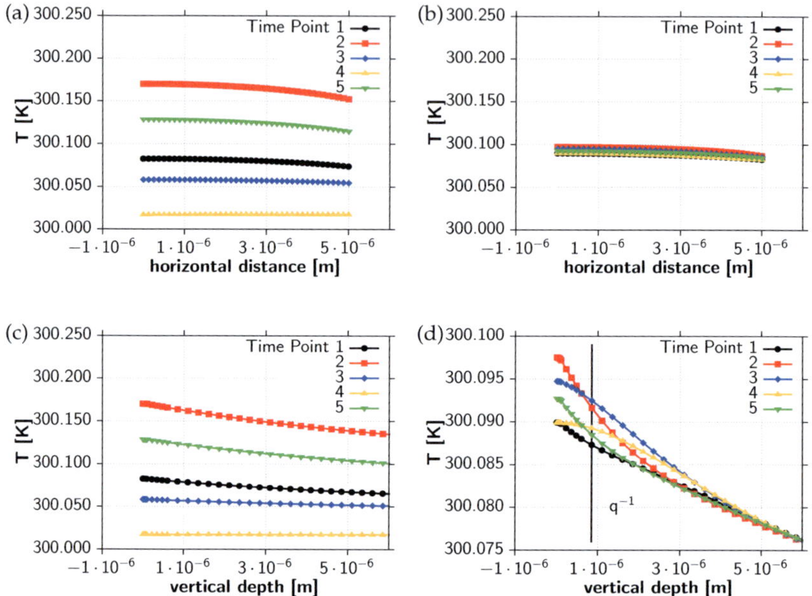

Figure 42: Temperature distribution along a (a),(b) horizontal and (c),(d) vertical cut. The system properties are: P_0/L = 5 W/m, Pt heater, 2b = 10 μm, a = 100 nm, MgO substrate, d_s = 1 mm, htc = 5·10^7 W/m^{2}K and f_c = 10 Hz in (a),(c) and f_c = 1 MHz in (b),(d).

4.1.1.5 Temperature dependent resistivity of the heater

The analytic descriptions for the temperature amplitude by Cahill [28] and Borca-Tasciuc et al. [22] assume a constant power per length. However, since the electric resistivity is coupled with temperature, the dissipated power changes even within a temperature cycle. This circumstance is incorporated with the linear temperature coefficient α into Eq.3.3. The BC's are like in the previous section.

A comparison between a fix and a temperature dependent resistivity is shown in Fig.43. In general two factors change the temperature amplitude.

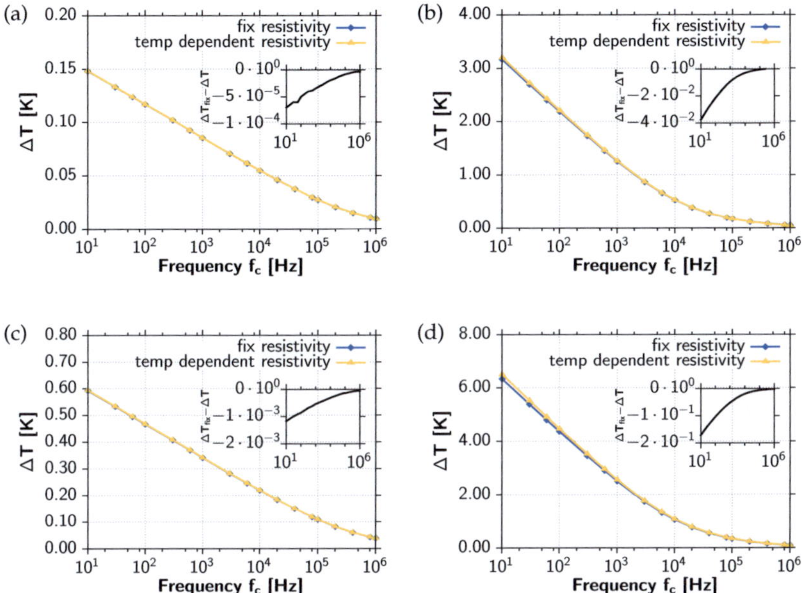

Figure 43: Influence of the temperature dependent electrical resistivity in the heater for a (a),(c) MgO and (b),(d) YSZ substrate. The system properties are: Pt heater, 2b = 10 μm, a = 100 nm, d_s = 1 mm, htc = $5 \cdot 10^7$ W/m²K and P/L = 5 W/m in (a),(b) and P/L = 20 W/m in (c),(d). The insets contain the difference in ΔT between a fix (ΔT_{fix}) and a temperature dependent resistivity (ΔT).

First, a poorly-conductive material leads to a higher ΔT than a highly-conductive material [cf. Fig.43(b) and 43(a)]. Thus, the difference is greater for poorly-conductive substrate materials. Second, a higher P/L increases the difference even more [cf. Fig.43(a) and 43(c)]. The biggest difference is consequently given in Fig.43(d), where the models differ on the first digit after decimal place. However, it has to be mentioned, that this represents the maximum in variation, since usually a 3ω-measurement takes place at lower P/L and especially for a ΔT around 1 K.

4.1.1.6 Three-dimensional heat conduction in the substrate

The analytic descriptions for the temperature amplitude by Cahill [28] and Borca-Tasciuc et al. [22] consider a two-dimensional model. To be valid, the length of the heater must be much longer than its width, so that $L \gg 2b$. However, even though this restriction is fulfilled, there is always heat conduction along the heater's length dimension [cf. Fig.44].

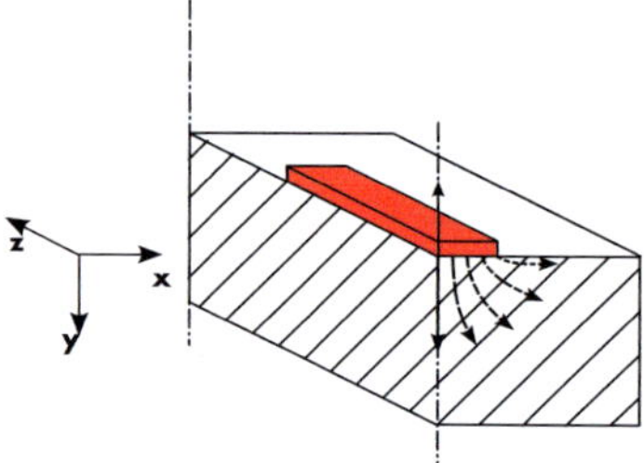

Figure 44: Considered system for the Finite Element simulation for three-dimensional heat conduction into the substrate.

In this section, every outer boundary is adiabatic except the substrate mount, where a htc is considered. The dissipated power is given by Eq.3.3. The resulting temperature amplitude in the heater is determined by Eq.3.13.

Fig.45 shows the three-dimensional temperature distribution for the time points ② and ④ [cf. Fig.40]. Like in Sec.4.1.1.4, the maximum and minimum temperatures are for f_c = 10 Hz always higher and lower than for f_c = 1 MHz. Therefore, compare Fig.45(a) and 45(c) versus Fig.45(b) and 45(d). Moreover, one can notice for both frequencies and time points a temperature drop along the heater towards the outer end. This circumstance is illustrated in Fig.46. Here, the spatial dependence of the temperature amplitude δT is plotted with the corresponding maximum and minimum temperature in the inset.

The cut is set along the z-axis below the center of the heater towards the outer edge of the substrate (z = 1 mm and L/2 = 0.5 mm). A comparison of a highly-, moderately- and poorly-conductive substrate material is given for f_c = 10 Hz in Fig.46(a), 46(b) and 46(c). Indicated by the green rectangle, the spatial temperature amplitude decreases towards the heater's end significantly earlier for highly-conductive than for poorly-conductive materials. On the contrary, a frequency increase keeps the spatial temperature amplitude over a longer distance constant. For this, compare Fig.46(b) versus 46(d).

A comparison between the two-dimensional models of Cahill of Eq.2.17 and Borca-Tasciuc of Eq.2.23 and a two- and three-dimensional Finite Element solution is given in Tab.3 for three different frequencies.

Solution ΔT	Cahill Eq.2.17	Borca-Tasciuc Eq.2.23	FE-2D	FE-3D
10 Hz	0.68859 K	0.68856 K	0.69507 K	0.67404 K
100 Hz	0.51094 K	0.51085 K	0.52034 K	0.51360 K
1 MHz	0.01861 K	0.01453 K	0.02265 K	0.02267 K

Table 3: Comparison of different solutions for a heater substrate system. The system properties are: P/L = 5 W/m, Pt heater, 2b = 10 μm, a = 100 nm, L/2 = 0.5 mm, STO substrate, d_s = 1 mm and htc = $5 \cdot 10^7$ W/m^{2}K.

According to Sec.4.1.1.1, the Finite Element solution is slightly above the analytic descriptions for two dimensions. However, a comparison of the two- and three-dimensional Finite Element solution results in a smaller ΔT for the three-dimensional analysis than for the two-dimensional case. This is due to a heat flow along the z-direction in the heater and the substrate. Since the heater has not the length of the substrate's width, the heater has its coolest cross-sectional cut at the end of its length. This circumstance is illustrated in Fig.45 and 46.

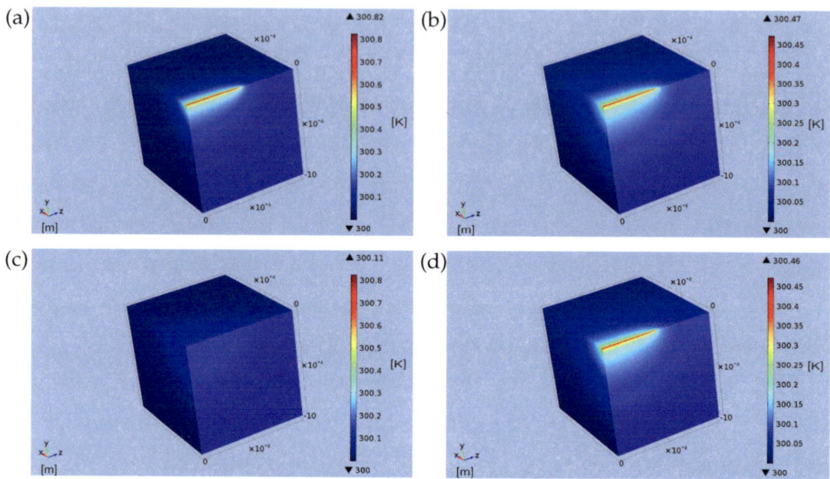

Figure 45: Visualization of the maximum and minimum temperatures within a certain temperature oscillation for (a)(c) f_c = 10 Hz and (b)(d) f_c = 1 MHz. The system properties are: P/L = 5 W/m, Pt heater, 2b = 10 μm, a = 100 nm, L/2 = 0.5 mm, STO substrate, d_s = 1 mm and htc = $5 \cdot 10^7$ W/m²K.

Hence, the mean temperature is lower for three-dimensional Finite Element solutions than for two-dimensional Finite Element solutions. Consequently, the experimentally measured temperature amplitude is smaller than the predicted ΔT from Finite Element solutions for two dimensions. Furthermore, it has to be mentioned that a cross-sectional cut into the three-dimensional heater substrate system, at the axis of symmetry (z = 0), results in the temperature amplitude ΔT of the two-dimensional model.

Thus, the heater's length must be sufficiently large and the measuring frequencies high enough to apply the two-dimensional models of Cahill and Borca-Tasciuc. But in any case, the determination of the thermal conductivity with a smaller ΔT results in a higher thermal conductivity. However, one has to distinguish

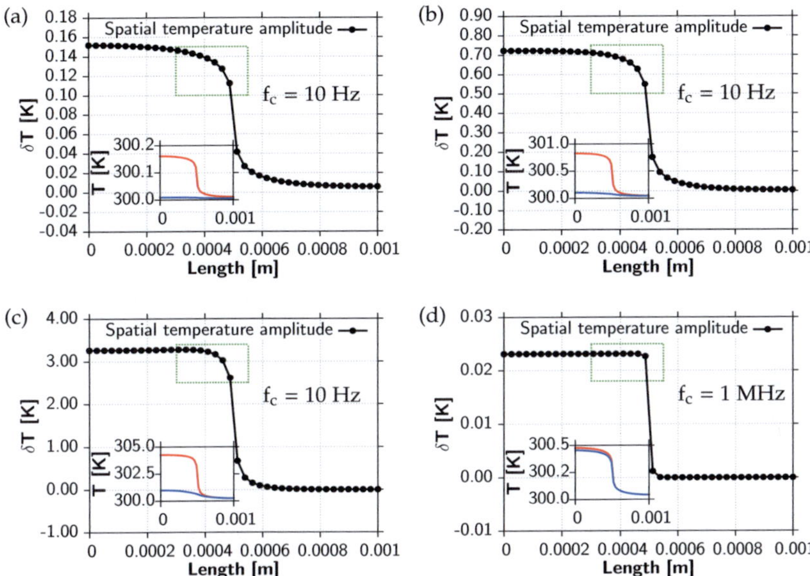

Figure 46: Influence of the heater's length on the spatial temperature amplitude δT along the z-axis for a (a) MgO, (b),(d) STO and (c) YSZ substrate. The system properties are: Pt heater, 2b = 10 µm, a = 100 nm, d_s = 1 mm and htc = $5 \cdot 10^7$ W/m^{2}K. The insets contain the maximum and minimum temperature along the z-axis.

between a 2-pad and a 4-pad measurement. The simulated case would represent a 2-pad measurement since the temperature amplitude in the heater is determined over the same length then the structure is heated. Thus also at the edges, where the temperature amplitude decreases [cf. 12]. In comparison, a 4-pad measurement takes the voltage within a defined length of a longer heater structure. Hence, this effect is minimized. Although, the influence of the three-dimensional heat flow is shown by Jacquot et al. [78], this study is necessary within this work to understand the general heat flow required to develop a heater substrate platform and three-dimensional membrane systems.

4.1.1.7 Radiation at the heater's surface

The previous sections consider exclusively heat transfer by heat conduction. However, while heat transfer by convection is avoided by performing a 3ω-measurement in vacuum, heat loss by radiation is always existent [cf. Sec.1.2]. Although Cahill [28] already examined the influence of radiation, the effect on the temperature amplitude is also included in this work. Thus we can study the influence of radiation for the material combinations which we use in this work, and the effect of radiation can be compared quantitatively to the other influence factors.

In general, heat loss by radiation is affected by four different factors. First, a broad heater, compared to a narrow heater, offers a greater surface for radiation to take place. Second, the more heat is conducted through the substrate, the less is going to be radiated. Thus, the thermal properties of the substrate limit indirectly the radiation losses. Third, heat loss by radiation is non-linear and changes with $\sigma_b e(T^4 - T_0^4)$ [cf. 35]. Hence, the ambient temperature T_0 has crucial impact onto the radiation losses. Here, σ_b is the Bolzmann constant and e is the emissivity of the materials surface.[35] To study the influence of radiation loss on the temperature amplitude, we consider a linearized approximation at the outermost boundary [cf. Fig.34(c)] so that

$$\Gamma_2: \qquad \kappa_{s/H}\frac{\partial T_{s/H}}{\partial n} - \sigma_b e T_0^3 \cdot (T_{s/H} - T_0) = 0. \qquad (4.6)$$

This linearized approximation is just appropriate for small temperature changes [35, p. 18], like it is in a 3ω-measurement. Fourth, the applied power per length defines basically the occurring temperatures.

---

[35]Find App.B. for detailed information about the Bolzmann constant and the emissivity.

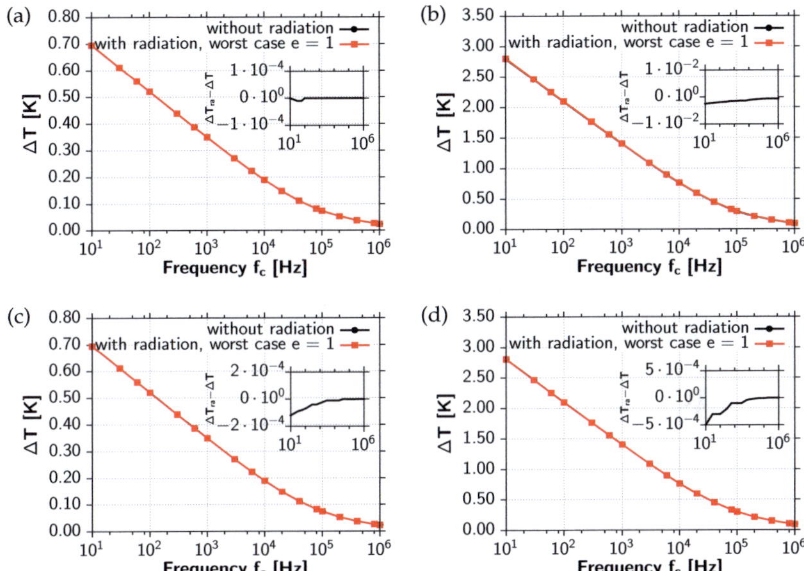

Figure 47: Influence of heat loss by radiation at the heater's boundaries for an ambient temperature of (a),(b) 300 K and (c),(d) 1300 K. The system properties are: Pt heater, 2b = 10 µm, a = 100 nm, STO substrate, d_s = 1 mm, htc = $5 \cdot 10^7$ W/m^{2}K and the emissivity is assumed to be e = 1. The applied power is P/L = 5 W/m in (a),(c) and P/L = 20 W/m in (b),(d). The insets contain the $\Delta T_{ra} - \Delta T$ difference between a system taking into account radiation (ΔT_{ra}) and without (ΔT).

Consequently, this interacts with the non-linear changes of T^4.

Boundary Γ_1 and Γ_3 are adiabatic and Γ_4 considers the substrate mount, where a htc is taken into account [cf. Fig.34(c)]. The dissipated power is given by Eq.3.3. Since the heater's width and the substrate material are quasi-fixed system quantities, the influence of ambient temperature and applied power per length are studied in Fig.47. To maximize the influence of radiation, the emissivity is set to e = 1.

While the difference in temperature amplitude ($\Delta T_{ra} - \Delta T$) is almost nonexistent for room temperature [cf. Fig.47(a) and 47(b)], there is a slight difference for high temperatures [cf. Fig.47(c) and 47(d)]. However, even for a high power per length and high temperature, the difference in temperature amplitude is first observed at the 4th decimal place. **Consequently, heat loss by radiation can be neglected for classic measuring configurations.**

4.1.1.8 Surface roughness

Also Half-Heusler alloys are nowadays subject of research for thermoelectrics. While the surface of STO, PCMO and MgO is rather flat, special Half-Heusler materials develop a surface roughness on the order of several nanometers [cf. 80, 81]. If the heater has the surface's shape, different heater width could occur. Fig.48 illustrates these circumstances.

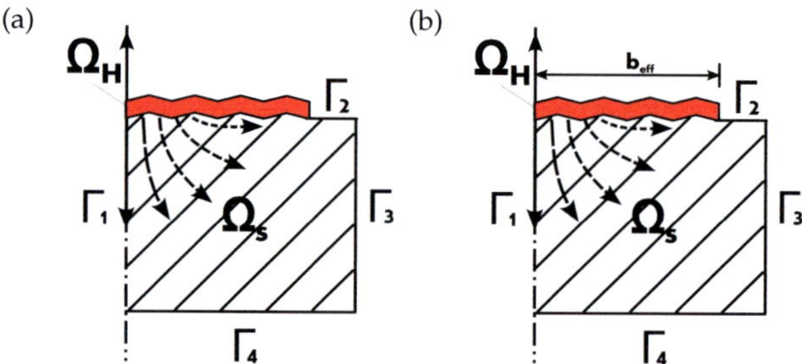

Figure 48: Considered systems for the Finite Element simulation with (a) surface roughness and (b) an effective heater width.

Here, we investigate the influence of surface roughness for a peak height of 20 nm on a 200 nm base. The boundaries Γ_{1-3} are adiabatic and Γ_4 takes into account a htc. The dissipated power is given by Eq.3.3. We study the difference between three different shapes of the heater.

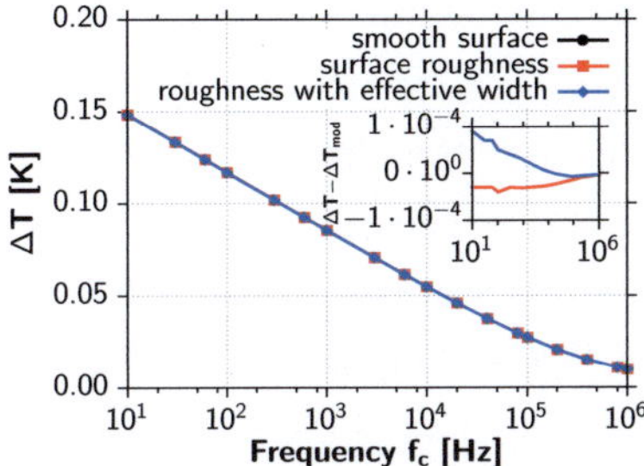

Figure 49: Influence of surface roughness on the cross-sectional cut of the heater. A waved surface of 20 nm peaks on a 200 nm base is realized. The system properties are: Pt heater, $2b = 10$ µm, $a = 100$ nm, MgO substrate, $d_s = 1$ mm and htc $= 5 \cdot 10^7$ W/m²K. The insets contain the difference between a flat heater (ΔT) and a system taking into account modifications of the heater due to a surface roughness (ΔT_{mod}).

The first shape is a rectangle with a smooth surface. The second shape has the same heater width than the rectangle with the smooth surface but exhibits a surface roughness. The third shape considers the same path length of the heater than the first shape, however exhibits a surface roughness too.

A comparison between these models is given in Fig.49. Here, the inset shows a difference of the solutions ($\Delta T - \Delta T_{mod}$) on 4th decimal place. **Thus, a 3ω-measurement is almost unaffected by such small variations of the heater geometry.** However, it has to be mentioned that the present Finite Element simulations neglect interface resistances. But, if the heater is not perfectly connected to the material below, interface resistances could become crucial.

4.1.1.9 Skin effect in the heater

Since the 3ω-method is an alternating current technique, the current density distribution has a tendency towards the boundary of the cross-sectional cut of the heater [cf. Sec.3.1.2]. Apart from material parameters, this tendency depends on the frequency of the current f_c. In general, the characteristic skin depth can be calculated with Eq.A.324. Applying this formula results in a skin depth of $\delta_{skin} = 0.1638$ m and $\delta_{skin} = 0.0001638$ m for $f_c = 10$ Hz and $f_c = 1$ MHz respectively. Obviously, this spatial distance exceeds the heater's width $2b = 10$ µm and height $a = 100$ nm by far, and thus almost no difference should occur in current density distribution along the heater's spatial dimension. However, since we focus even on the smallest changes in temperature amplitude, this effect is studied here to examine the influence on ΔT.

The boundaries Γ_{1-3} are adiabatic and Γ_4 takes into account a htc. The dissipated power is given by Eq.3.3. However, in this equation, the current density is represented by the effective current density j_{eff} of Eq.3.45.

The new occurring current density distribution is shown in Fig.50 for two different heater widths and three different frequencies. While there is almost no gradient observable for the heater width of $2b = 10$ µm at $f_c = 10$ kHz in Fig.50(a), the distribution changes slightly towards the outer edge of the heater for $f_c = 1$ MHz in Fig.50(c) and 50(e). On the contrary, a heater width of $2b = 50$ µm also comprises a different distribution at $f_c = 10$ kHz [cf. Fig.50(b)]. Apparently, the distribution does not vary in the heater's height for both given configurations [cf. 83]. Even if a tremendous heater with the dimensions of $a = 2b = 50$ µm would be applied for measurements, almost no gradient is observable in the density distribution at $f_c = 10$ Hz [cf. Fig.51(a)].

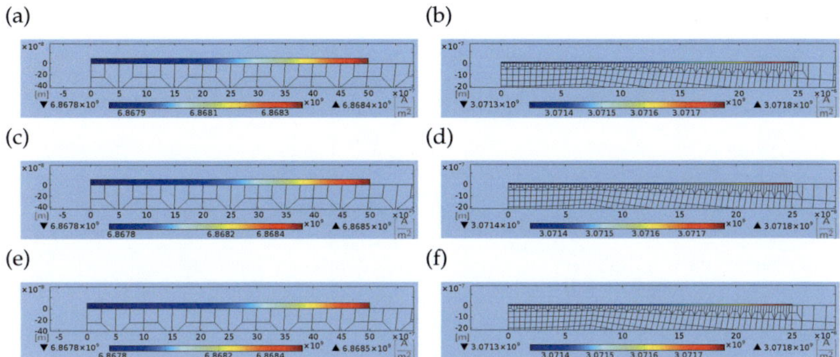

Figure 50: Visualization of the current distribution for a heater width of 2b = 10 µm (a),(c),(e) and 2b = 50 µm (b),(d),(f) for 10 kHz, 100 kHz and 1 MHz respectively. The system properties are: P/L = 5 W/m, Pt heater, a = 100 nm, STO substrate, d_s = 1 mm and htc = $5 \cdot 10^7$ W/m^{2}K.

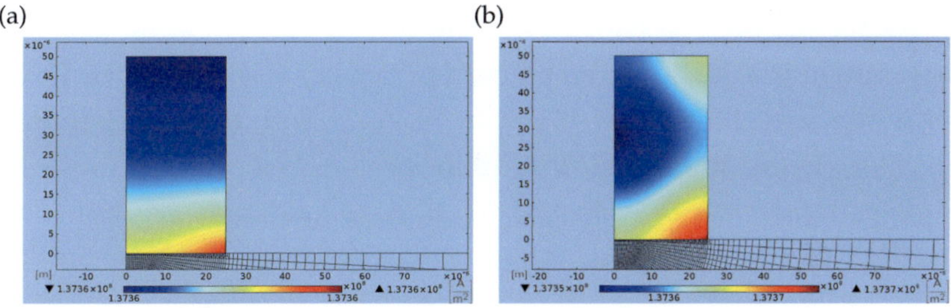

Figure 51: Visualization of the current distribution in a squared heater for (a) 10 Hz and (b) 1 MHz. The system properties are: P/L = 5 W/m, Pt heater, dimensions of a = 2b = 50 µm, STO substrate, d_s = 1 mm and htc = $5 \cdot 10^7$ W/m^{2}K.

On the contrary, at f_c = 1 MHz, a gradient would occur in the direction of height and width [cf. Fig.51(b)] [cf. 64]. However, it has to be mentioned that these heater dimensions are unlikely to be used for a 3ω-measurement.

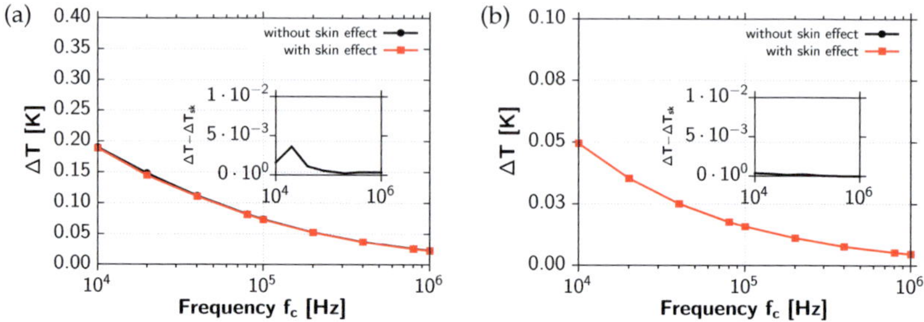

Figure 52: Influence of the skin effect for a (a) narrow and a (b) broad heater. The system properties are: $P/L = 5$ W/m, Pt heater, $a = 100$ nm, STO substrate, $d_s = 1$ mm and htc $= 5 \cdot 10^7$ W/m^{2}K. The insets contain the $\Delta T - \Delta T_{sk}$ difference between a system with an equally distributed current density and a system taking into account the skin effect in the heater.

The influence on the temperature amplitude is shown in Fig.52. The difference between a changed heating due to the skin effect and the assumption of equal distributed current density is shown in the inset for a heater width of $2b = 10$ µm in Fig.52(a) and for $2b = 50$ µm in Fig.52(b). **Even though a different density distribution exists towards higher frequencies, the effect of the changed heating is negligible in the high frequency regime.**

4.1.1.10 Thermal expansion and stresses

The measurement principle of the 3ω-method is based on heat production in the heater and dissipation into the investigated material. Analytic descriptions of Cahill and Borca-Tasciuc consider exclusively fixed geometries. However, a temperature change in a material is often coupled with a mechanical deformation. Depending on the thermal expansion coefficients of the respective materials, the spatial dimension of the heater is subject to change.[36]

---

[36] Although we assume small deformations [cf. Sec.3.1.3], the influence of mechanical expansion on the temperature amplitude is investigated here.

Hence, the power input is given through a continuously changing cross-sectional cut of the heater and thus, the current density varies throughout a temperature cycle. To take into account the mechanical displacement, the current density is calculated for every time step by

$$j = \frac{I_0}{(a + u_{y(a)}) \cdot (b + u_{x(b)})} ,$$ (4.7)

where, $u_{y(a)}$ and $u_{x(b)}$ are the mechanical displacements in height and width at the outer edge of the heater. Hence, the heat source term of Eq.3.3 writes now as

$$p = \varrho_0 \cdot (1 + \alpha \cdot (T - T_0)) \cdot \left(\frac{I_0}{(a + u_{y(a)}) \cdot (b + u_{x(b)})}\right)^2 \cdot \frac{1}{2} \cdot (1 + \cos(2\omega t)).$$ (4.8)

This time-varying process is taken into account for two representative substrate materials (STO and YSZ) in Fig.53(a) and 53(b) respectively. The thermal conditions are the same as in the previous section. The conditions of the mechanical field are given in Sec.3.1.3. Since the thermal expansion coefficient (1/ K) depends on the relatively changing temperature, this effect can be maximized always by applying a higher power per length. Assuming the materials are in perfect mechanical contact, the difference in ΔT is at its maximum only at the 3rd decimal place for low frequencies. This difference levels off for higher frequencies. For the previously given study, the mechanical behavior is visualized in Fig.54 for the maximum temperature at time point ② in Fig.40. According to the thermal and mechanical boundary conditions of Sec.3, the maximum displacements in x-direction are at the upper right edge of the cross-sectional cut of the system [cf. Fig.54(a)].

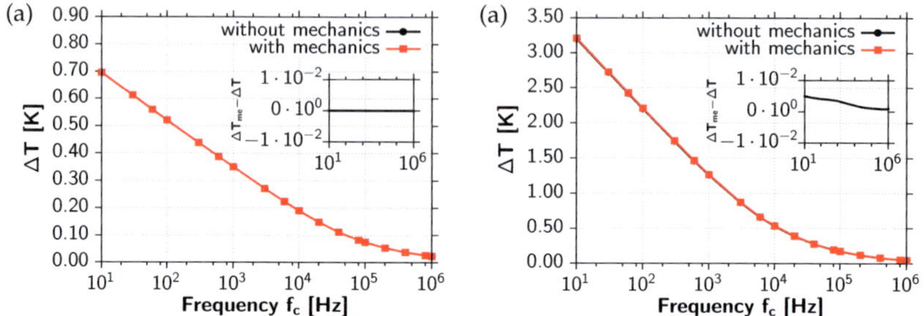

Figure 53: Influence of mechanical expansion for a (a) STO and (b) YSZ substrate material. The system properties are: P/L = 5 W/m, Pt heater, 2b = 10 μm, a = 100 nm, d_s = 1 mm and htc = $5 \cdot 10^7$ W/m^{2}K. The insets contain the $\Delta T_{me} - \Delta T$ difference between a system including mechanics and a system without mechanics.

The maximum displacement in y-direction is given at the upper left edge of the system, thus at the top center of the heater. In both cases, the maximum displacement is around a hundredth of the heater's height. The normal stresses in x- and y-direction are shown in Fig.54(c) and 54(d), and the shear stress is shown in Fig.54(e). The occurring stresses are calculated by [8, p. 54]

$$\sigma_x = 2Gu_{x_x} + \frac{2Gv}{(1-2v)}\left(u_{x_x} + u_{y_y}\right) - \frac{E_\sigma}{1-2v}\beta(T - T_0),\tag{4.9}$$

$$\sigma_y = 2Gu_{y_y} + \frac{2Gv}{(1-2v)}\left(u_{x_x} + u_{y_y}\right) - \frac{E_\sigma}{1-2v}\beta(T - T_0),\tag{4.10}$$

and the shear stress is

$$\tau_{xy} = G \cdot \left(u_{x_y} + u_{y_x}\right).\tag{4.11}$$

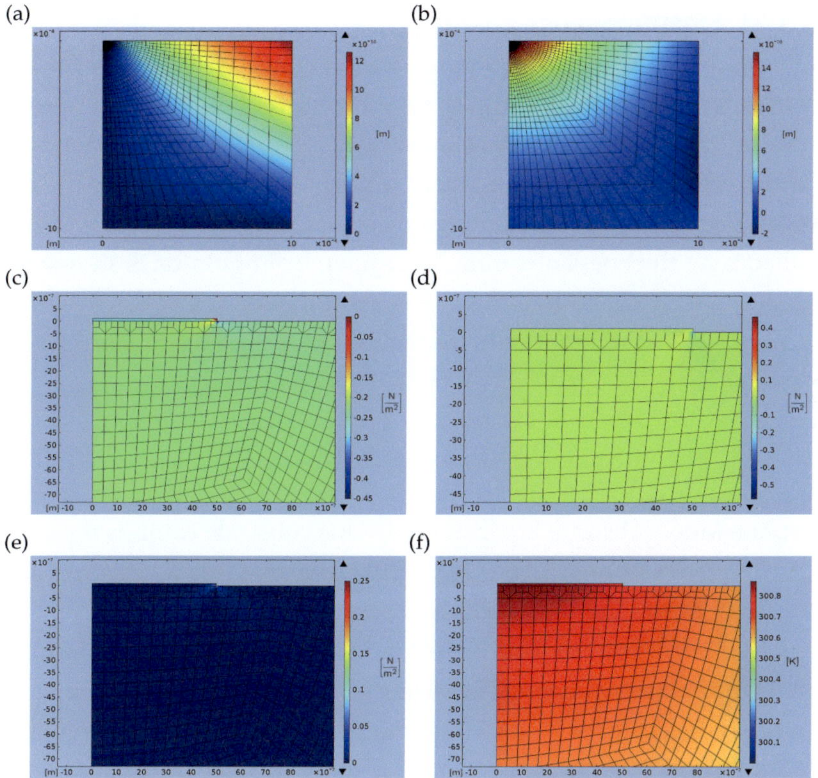

Figure 54: Visualization of the mechanical displacements and resulting stresses for the maximum temperature within a certain oscillation [cf. Fig.40]. (a) is the displacement in x-direction, (b) is the displacement in y-direction, (c) is the stress in x-direction, (d) is the stress in y-direction and (e) is the shear stress in the x-y-plane. For the time point with maximum temperature, the distribution is illustrated in (f). The system properties are: $f_c = 10$ Hz, $P/L = 5$ W/m, Pt heater, $2b = 10$ μm, $a = 100$ nm, STO substrate, $d_s = 1$ mm and htc $= 5 \cdot 10^7$ W/m^{2}K.

For each, the maximum stress occurs again at the center of the heater. The absolute stress values are < 3 N/m^2 which is several orders of magnitude below critical stresses [cf. 6]. The corresponding temperature distribution is shown in Fig.54(f).

4.1.1.11 Eigenfrequency analysis

In this section, the eigenfrequencies are examined. Here, we consider only heater substrate configurations where a spatial platinum heater is placed on top the respective substrates. The platinum heater has a width of 10 μm and a height of 100 nm. Corresponding to the previous sections, the oxide substrate materials MgO, STO and YSZ are studied for a thickness of 1 mm. The system configuraion is shown in Fig.34(c). While the boundaries $\Gamma_2 - \Gamma_4$ are fully free to move, we set the mechanical displacement only for the axis of symmetry in x-direction to zero so that

$$\Gamma_1 : u_x = 0. \tag{4.12}$$

First, the MgO substrate is considered. This substrate exhibits the first eigenfrequency at $\omega = 1.56\,\text{MHz}$. However, this frequency must be compared to the thermal excitation and not to the excitation of the applied electric current. In a 3ω-measurement, the thermal excitation equals the heating frequency 2ω. Hence, the obtained eigenfrequency is already reached for $\omega/2$. Thus, the first critical frequency for the MgO substrate is 0.780 MHz. The corresponding eigenform describes a flapping of the substrate symmetric to boundary Γ_1. This eigenform is shown in Fig.55. Due to the boundary conditions, this is the first eigenform for every substrate material. The STO substrate exhibits the first eigenfrequency at $\omega = 1.21\,\text{MHz}$ and thus the first critical frequency for the STO substrate is 0.605 MHz. The YSZ substrate exhibits the first eigenfrequency at $\omega = 1.02\,\text{MHz}$ and thus the first critical frequency for the YSZ substrate is 0.510 MHz. However, in a 3ω-measurement the substrate is always placed on a mount at boundary Γ_4 and hence this boundary can not move as freely as we assumed it.

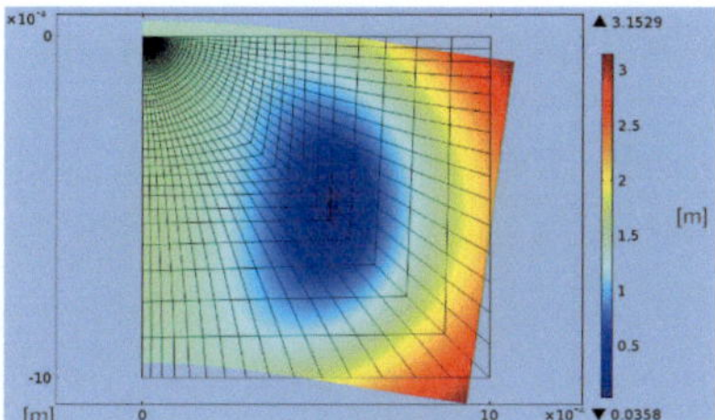

Figure 55: Visualization of the eigenmode at the first eigenfrequency $\omega = 1.56$ MHz for an MgO substrate.

In any case, the given eigenfrequencies indicate that for these substrates, 3ω-measurements can be performed up to 0.5 MHz without a resonance catastrophe.

4.1.2 Heat source on layer-substrate materials

This section examines the general system behavior for a heat source on layer-substrate materials. First, monolayer configurations are studied [cf. Fig.56(a)] and the solutions for the temperature amplitude are compared to the descriptions of Lee and Cahill [99] and Borca-Tasciuc et al. [22]. The potential for anisotropic thermal conductivity determination is examined for highly-conductive and poorly-conductive substrate materials. Afterwards, multilayer configurations are studied [cf. Fig.56(b) and 56(c)] for the general system behavior and the influence of different layer thickness and sequence on the temperature amplitude.

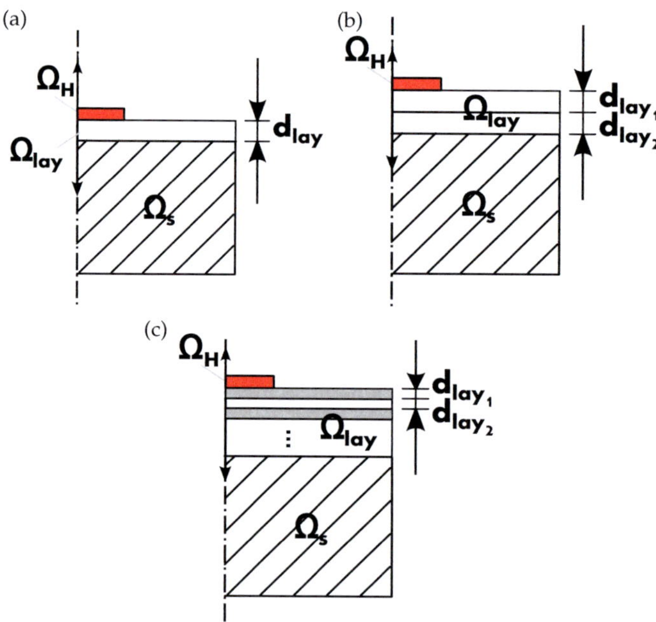

Figure 56: Considered layered systems for the Finite Element simulations: (a) Heat source on one layer; (b) on two layers; (c) on a multilayer stack.

In this section, only temperature fields are considered. The boundaries Γ_{1-3} are adiabatic and Γ_4 takes into account a htc [cf. Fig.27]. The power input is given by Eq.3.3.

4.1.2.1 Monolayer configurations

In general, one must distinguish between two different layer substrate configurations. On one hand, there can be a relatively poorly-conductive layer material on top of a highly-conductive substrate material and on the other hand, the opposite configuration. First, the classic configuration with a relatively poorly-conductive layer material is considered. Here, the 3ω-method is especially convenient to determine the cross-plane thermal conductivity. However, even though a 3ω-measurement is performed within the boundaries of the offset model [cf. Eq.2.30], the temperature amplitude is affected by the in-plane heat spread. Therefore, Cahill and Lee [33] proposed a correction value to take this fact into account and revised the offset model. Conversely, the solution of Borca-Tasciuc et al. [22] takes this fact analytically into account. Moreover, this solution accounts for thermal properties of the heater to describe the system behavior more precisely in the high frequency regime.

A comparison between the solutions of the different models is given in Fig.57(a). First, the low frequency regime is considered until a frequency of $f_c = 10$ kHz. Up to this frequency, the offset model of Eq.2.31 is valid in general. In this regime, the corrected offset model of Cahill [Eq.2.31 including Eq.2.33], the solutions of Borca-Tasciuc [Eq.2.24 and Eq.2.23], and the Finite Element solution are almost coincident. Only the original offset model of Cahill [Eq.2.31], without the effective heater width differs slightly.

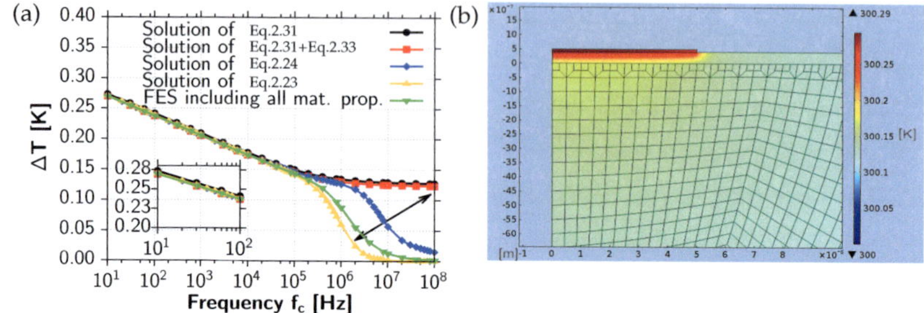

Figure 57: (a) Influence of in-plane heat spread and of heater properties in the high frequency regime. (b) The corresponding temperature distribution illustrates the profound temperature drop below the heater at $f_c = 10$ Hz. The system properties are: $P/L = 5$ W/m, Pt heater, $2b = 10$ µm, $a = 100$ nm, PCMO layer, $d_{lay} = 400$ nm, isotropic layer, MgO substrate, $d_s = 1$ mm and htc $= 5 \cdot 10^7$ W/m²K.

Since the slight in-plane heat spread generally lowers the temperature amplitude, this ΔT-line is slightly above the other four solutions. The inset contains a scope of the temperature amplitude solutions between 10 Hz and 100 Hz. Here the analytic solutions agree even better with the Finite Element solution than for the substrate validation in Sec.4.1.1.1.

On the contrary, a comparison in the high frequency regime reveals significant differences. Since the offset model treats the layer as a one-dimensional resistor and thus does not include the frequency dependence of the layer into the solution, the offset model solutions must tend to a remaining temperature amplitude even for an infinite f_c [cf. Fig.24]. In contrast, the solutions of Borca-Tasciuc of Eq.2.24 and Eq.2.23 decrease significantly towards $f_c = 100$ MHz. The solution of Eq.2.24, without the heater properties, remains for higher frequencies, at considerably larger temperature amplitudes than the solution of Eq.2.23 and the Finite Element solution. The significant decrease of ΔT at $f_c = 10$ MHz is due to the thermal penetration depth.

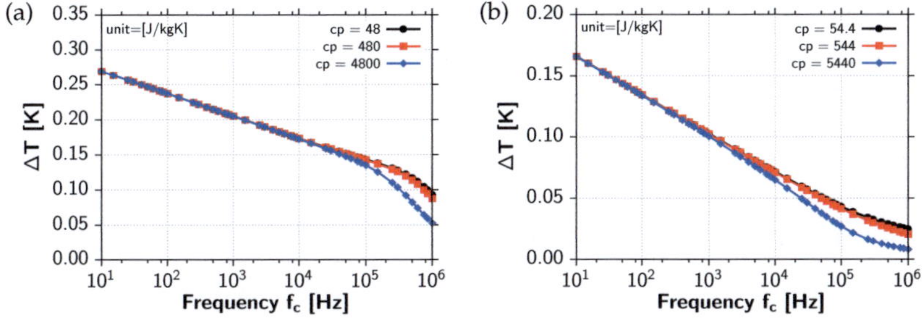

Figure 58: Influence of a thermal mass variation on the temperature amplitude for an (a) PCMO layer and an (b) STO layer. The system properties are: P/L = 5 W/m, Pt heater, a = 100 nm, 2b = 10 µm, d_{lay} = 400 nm, MgO substrate, d_s = 1 mm, htc = $5 \cdot 10^7$ W/m^{2}K.

Here, the penetration depth is on the order of the layer's thickness. However, if the thermal mass ($\rho \cdot c_p$) of the heater is considered additionally in Borca-Tasciuc's solution [Eq.2.23], the ΔT-line decreases already at frequencies around f_c = 300 kHz. This is due to the thermal wave, which interacts now significantly with the thermal properties of the layer and the heater. The Finite Element solution considers in contrast to the analytic descriptions all material properties of the heater. Since only the solution of Eq.2.23 takes into account thermal porperties of the heater, this ΔT-line is the closest to the Finite Element solution. Although the solutions of Borca-Tasciuc can be seen as bounds from above and below, the Finite Element Model helps to understand the ΔT evolution in the high frequency regime. Furthermore, it demonstrates the necessity of Finite Element simulations in terms of high precision analysis for new measuring methods. Moreover, this result supports the assumption of neglecting the thermal properties of the heater in the low frequency regime for classic layer substrate geometries and configurations.

On the contrary, the question arises how a thermal mass variation of the layer material takes an influence on the temperature amplitude. To study the influence on ΔT, the original thermal mass of an PCMO and STO layer is varied by a factor of ten. Hence three different cases arise for both materials. The resulting temperature amplitudes are shown in Fig.58. **Evidently, the thermal mass of the layer material influences the temperature amplitude exclusively for higher frequencies.**
The actual temperature distribution for the 400 nm thick PCMO layer on top the MgO substrate is shown in Fig.57(b) for the time point ② in Fig.40. Even though a slight in-plane heat spread is present in the layer, a strong pronounced vertical temperature drop below the heater exists. Based on this temperature distribution, the offset model is justified.
On one hand, the in-plane heat spread is undesired for the cross-plane thermal conductivity determination. One the other hand, the in-plane heat spread can be used to determine the in-plane thermal conductivity. As mentioned in Sec.2.2.2, some groups applied narrow heaters to determine the in-plane thermal conductivity for relatively thick and highly-conductive materials by this in-plane heat spread. However, the materials investigated in this work are thin and poorly-conductive. The sensitivity on the temperature amplitude is shown in Fig.59 for tremendous varying cross- and in-plane thermal conductivities. In this work, sensitivity is defined as significant derivation from the isotropic case of the ΔT-line. This includes different slopes and changing axis-intercepts at various frequencies. Numerically, changes can be observed easily, but the value changes in the temperature amplitude must significantly exceed the measurement error in experiments. The measurement error of the experimental setup, used by our project partners, is determined to be around 5% [cf. 164].

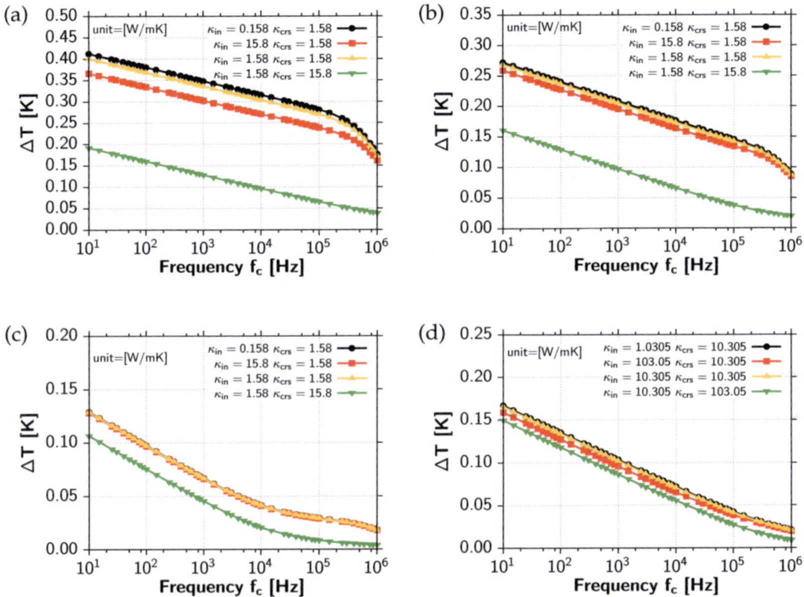

Figure 59: Sensitivity for the anisotropic thermal transport. Changes in ΔT for a PCMO layer including a (a) narrow $2b = 5\ \mu m$, (b) intermediate $2b = 10\ \mu m$ and (c) broad $2b = 50\ \mu m$ heater width. Changes in ΔT for a STO layer including an (d) intermediate heater width for $2b = 10\ \mu m$. The system properties are: $P/L = 5\ W/m$, Pt heater, $a = 100\ nm$, $d_{lay} = 400\ nm$, MgO substrate, $d_s = 1\ mm$, htc $= 5 \cdot 10^7\ W/m^2K$.

In Fig.59(a), 59(b) and 59(c), different heater widths are applied onto a thin poorly-conductive PCMO layer. To investigate the sensitivity for the anisotropic thermal conductivities, the isotropic case is varied by a factor of ten. Hence five different cases arise. The case of extremely poorly-conductive cross-plane thermal conductivity is neglected in Fig.59.

Except for moderately- and highly-conductive cross-plane thermal conductivities, only small differences occur in the ΔT-lines for a narrow, 'normal' and broad heater width [cf. Fig.59(a), 59(b) and 59(c)]. Also the anisotropic variation of an moderately-conductive STO layer results only in small differences in the ΔT-lines [cf. Fig.59(d)]. Moreover, here the absolute values of the temperature amplitudes are already small and the occurring differences in the ΔT-lines compared to the tremendous variations of the thermal conductivities are even smaller.

For classic systems with a highly-conductive substrate, one can distinguish clearly between the influence of thermal conductivity and thermal mass of the layer on the temperature amplitude. Because here, a variation of the anisotropic thermal conductivity results in a constant slope until the thermal penetration depth is on the order of the layer. Moreover, since the slope is the same, but the axis intercept changes, the offset model can be applied in general, if the variations in ΔT are greater than the measurement accuracies. The ratio of the isotropic thermal conductivity of layer to substrate are for the PCMO-MgO system $\kappa_{\mathrm{lay}}/\kappa_{\mathrm{s}} \approx 0.03$ and for the STO-MgO system $\kappa_{\mathrm{lay}}/\kappa_{\mathrm{s}} \approx 0.18$. If a PCMO layer would be applied onto a STO substrate, the ratio would be $\kappa_{\mathrm{lay}}/\kappa_{\mathrm{s}} \approx 0.15$. In this case, the offset model could also be used to determine the cross-plane thermal conductivity.

The opposite layer substrate configuration, with a poorly-conductive substrate, results in a completely different system behavior. Because here, a comprehensive influence of the thermal conductivity and thermal mass, thus the thermal diffusivity, is present over the whole frequency regime. For these systems, the slope and axis intercept change continuously. Here, a classic linear regime does not exist. This system behavior is shown in Fig.60(a) for an STO layer and in Fig.60(b) for an PCMO layer for various layer thicknesses.

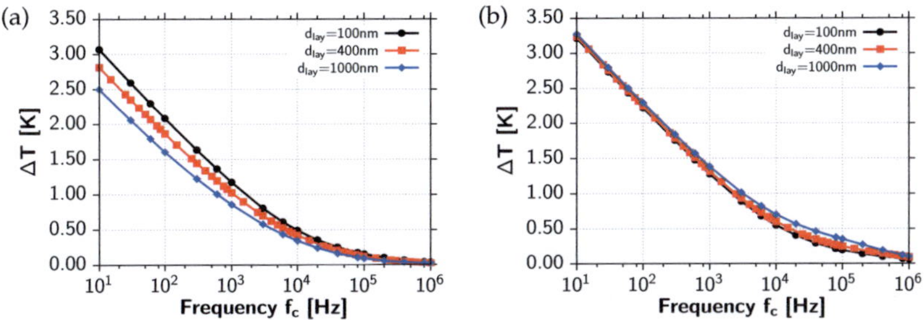

Figure 60: Non-offest system behaviour for a (a) STO and (b) PCMO layer at various heights. The system properties are: P/L = 5 W/m, Pt heater, a = 100 nm, 2b = 10 µm, YSZ substrate, d_s = 1 mm, htc = $5 \cdot 10^7$ W/m^{2}K.

Hence, there is also no clearly defined offset or downshift due to the STO and PCMO layer in the ΔT-line. **As consequence, the cross-plane thermal conductivity cannot be determined by the solution of Cahill and Lee [33] under these circumstances.** However, since we are especially interested in the anisotropic thermal conductivity, the poorly-conductive substrate is further investigated. The sensitivity for the anisotropic thermal conduction is shown in Fig.61. Although we investigate tremendous variations in the anisotropic thermal conductivity, just two cases vary significantly from the isotropic case. The highly-conductive in-plane case for STO [Fig.61(a)] and the poorly-conductive cross-plane case for PCMO [cf. Fig.61(b)]. Even though the influence on the temperature amplitude of the in-plane thermal conductivity can be increased by a narrow heater, the YSZ substrate is also rather insensitive for poorly-conductive in-plane thermal conductivities. Moreover, the application of poorly-conductive substrates would require also measurements with a highly-conductive substrate to determine also the cross-plane thermal conductivity first.

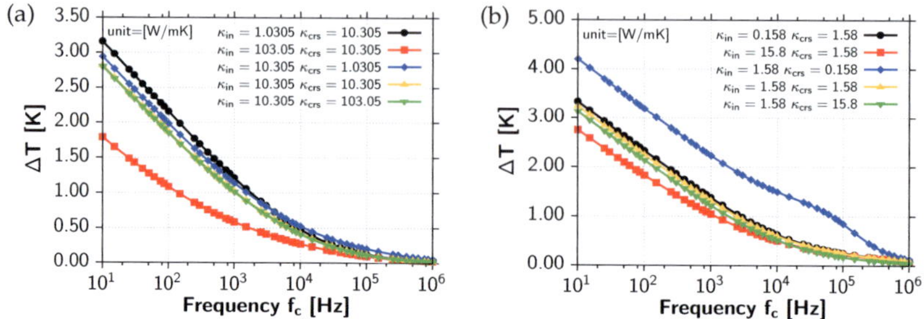

Figure 61: Sensitivity for the anisotropic thermal conduction for a (a) STO and (b) PCMO
layer. The system properties are: P/L = 5 W/m, Pt heater, a = 100 nm, 2b =
10 µm, d_{lay} = 400 nm, YSZ substrate, d_s = 1 mm, htc = $5 \cdot 10^7$ W/m²K.

By this process, several unknown conditions are introduced by each
measurement so that the determination of the thermal conductivity
is adversely affected. Consequently, this work aspires to determine
the in-plane and cross-plane thermal conductivity within one
measurement for even thinner layers. Therefore, new measurement
configurations are still necessary. For this reason, this work focuses
also on the bottom electrode and membrane geometries in section
Sec.4.2 and Sec.4.3, respectively.

4.1.2.2 Multilayer configurations

Multilayer systems are configurations with at least two individual
layers. The term multilayer is not specific about the layer's thickness.
However, new thermoelectric material properties are expected at
length scales around several nanometers for each layer thickness. Here,
especially a new emergent thermal conductivity arises. But since the
joint project is also interested in layer thicknesses up to several hundred
nanometers, the influence of layer thickness and stacking sequence is
investigated here.

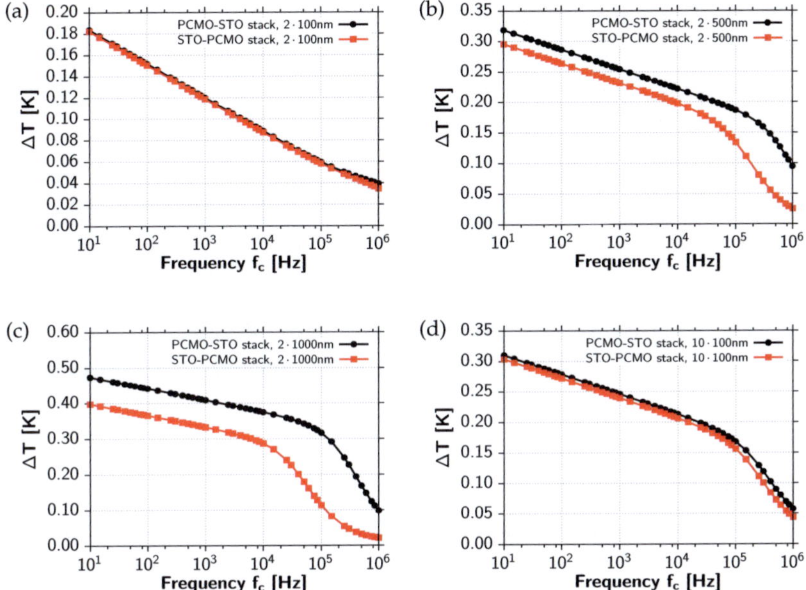

Figure 62: Influence of different layer ordering on the temperature amplitude. A variation of the top layer in a twin configuration is given for a single layer thickness of (a) 100 nm, (b) 500 nm and (c) 1000 nm. Influence of the layer ordering in a multilayer system on the temperature amplitude for a stack thickness of 1000 nm for (b) $2 \cdot 500$ nm and (d) for $10 \cdot 100$ nm. The system properties are: $P/L = 5$ W/m, Pt heater, $a = 100$ nm, $2b = 10$ μm, MgO substrate, $d_s = 1$ mm, htc $= 5 \cdot 10^7$ W/m^{2}K.

Therefore, different combinations are studied and shown in Fig.62. While the layer ordering is almost irrelevant for a stack thickness of 200 nm, the difference between the observed ΔT -lines is more pronounced towards a stack thickness of 2000 nm [cf. Fig.62(a), Fig.62(b) and 62(c)]. In the following, the different individual layer thickness is compared for a stack thickness of 1000 nm.

While an individual layer thickness of d_{lay} = 500 nm results in two clearly distinct ΔT-lines, the ΔT-lines are close to each other for an individual layer thickness of d_{lay} = 100 nm [cf. Fig.62(b) and Fig.62(d)]. Thus, the individual layer thickness should be as thin as possible to apply the offset model of Cahill. The reason for this system behavior is the in-plane heat spread in the layer below the heater. Shin et al. [144] also observed this upper layer influence. Without the in-plane heat spread, the ΔT-lines would be coincident until the thermal penetration depth is on the order of the stack thickness. In this case, the cross-plane thermal conductivity could be determined with the offset model without an influence of the layer ordering. However, a transition between an individual layer influence and a new representative material affects the temperature amplitude measurements differently. This transition depends besides each respective layer thickness also on the heater's width and the measuring frequency. **But, it can be concluded from Fig.62(d), that the in-plane heat spread from the upper layer can be neglected for a poorly-conductive stack with a total thickness of $d_{lay_{tot}} \leq (2b)/10$ and an individual layer thickness of $d_{lay} < 100$ nm.**
Thus, it is rather inappropriate to model each layer under these circumstances. Here, a new representative layer package can be regarded for the Finite Element simulations to represent the stack. Consequently, anisotropic thermal conductivity variations of a multilayer package are comparable with the monolayer studies of Sec.4.1.2.1.

4.2 Bottom electrode geometry

A possible new geometry for the 3ω-method is the bottom electrode geometry with the so-called heater substrate platform [cf. Fig.26].[37] A main advantage would be the improved reproducibility of the corresponding measurements, as there are equal heat transfer conditions to the substrate for each investigated material to be characterized. Before a heater substrate platform can be applied for nanostructured samples, one must understand the whole system behavior. Thus, the following section examines the general ΔT -line behavior for bulk-like materials first. Subsequently, full covering thin films are studied. Afterwards patterned multilayers are considered and compared to the classic geometry configuration.

Every subsequent section on the bottom electrode considers the same thermal boundary conditions. Except the substrate mount, every outer boundary is considered to be adiabatic. The substrate mount takes into account a htc to develop a realistic system behavior. The liberated power is given by Eq.3.3.

4.2.1 Bulk-like materials

In this section, three different geometry configurations are investigated. First, only thermal contact is given over the boundaries of the heater [cf. Fig.63(a)]. This geometry represents an ideal case to study the influence of either thermal conductivity or thermal mass variations on the temperature amplitude. The next step considers a more realistic case. Here, the placed material above the heater substrate platform also has spatial contact next to the heater [cf. Fig.63(b)].

---

[37]The bottom electrode concept is the basic idea of Ch. Jooss, Institute of Materials Physics, University of Göttingen.

At last, full thermal contact is established and compared to the spatial contact systems [cf. Fig.63(c)].

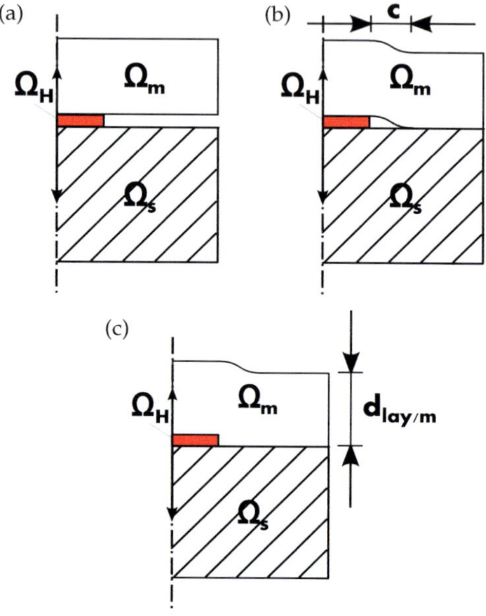

Figure 63: Considered systems for the Finite Element simulations for (a) a heat source between bulk-like materials, (b) a heat source between bulk-like materials with spatial contact and (c) a heat source between bulk-like materials without a gap.

In general, the thermal diffusivity $D = \kappa/(\rho \cdot c_p)$ can be varied by either changing the thermal conductivity or the thermal mass. Consequently, a variation of both could lead to the same diffusivity. However, both variations have significantly different influences on the measured temperature amplitude. Fig.64 shows the different influences. **While a thermal conductivity variation of the material above the heater substrate platform tremendously changes the slope of the ΔT -lines [cf. Fig.64(a)], a thermal mass variation only changes the axis intercept but the slope remains the same [cf. Fig.64(b)].**

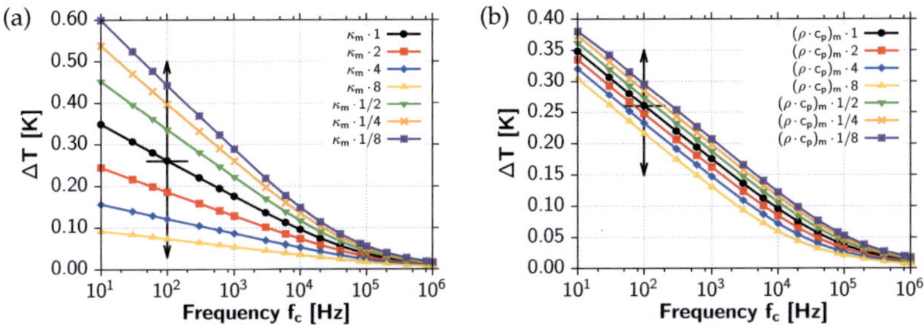

Figure 64: Influence of (a) thermal conductivity variation and (b) thermal mass variation on the temperature amplitude. The system properties are: P/L = 5 W/m, Pt heater, a = 100 nm, 2b = 10 µm, STO substrate material as origin, d_s = d_m = 1 mm, htc = $5 \cdot 10^7$ W/m²K.

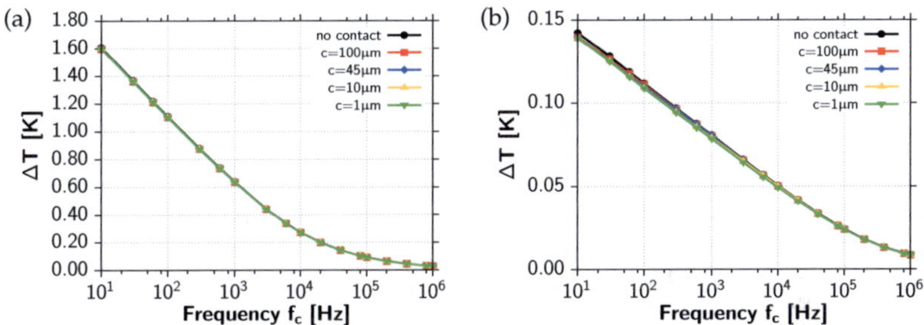

Figure 65: Influence of spatial contact between the substrate material YSZ and placed material (a) YSZ and (b) MgO on the temperature amplitude. The system properties are: P/L = 5 W/m, Pt heater, a = 100 nm, 2b = 10 µm, d_s = d_m = 1 mm, htc = $5 \cdot 10^7$ W/m²K.

Apparently, for both variations, the ΔT-lines must tend to zero temperature amplitude for higher frequencies. Hence, one cannot distinguish between the different influence factors here. For these ideal systems, one could determine the thermal conductivity of the sample on top of the heater substrate structure through the constant slope in the linear regime by the proposed formula of Eq.2.34 given in Sec.2.3.1.

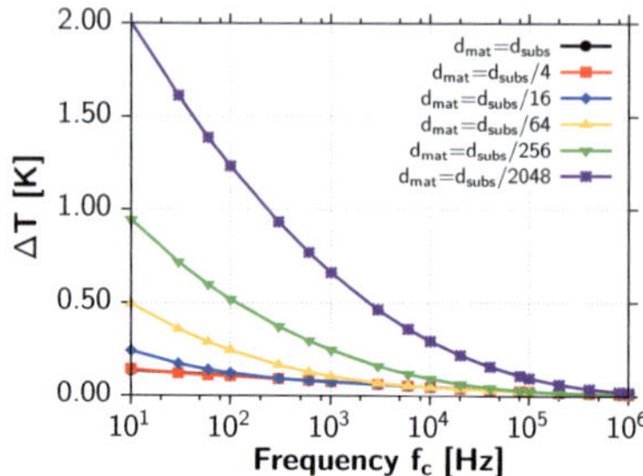

Figure 66: Influence on the temperature amplitude for decreasing material thickness on top. The system properties are: P/L = 5 W/m, Pt heater, a = 100 nm, 2b = 10 μm, MgO material on top, d_s = 1 mm, htc = $5 \cdot 10^7$ W/m²K.

However, these ideal systems are rather difficult to manufacture. In most experiments, the sample on top would probably also be in thermal contact with the substrate material [cf. Fig.63(b)]. Since the thermal conductivity of both materials is contained in the slope of the ΔT -line, the maximum difference is given for a high contrast in thermal conductivities. In Fig.65(a), the same poorly-conductive substrate material is taken as sample material on top so that $\kappa_m/\kappa_s = 1$. In this case, the obtained ΔT -lines are coincident for various contact distances c. On the contrary, for a contrast of $\kappa_m/\kappa_s \approx 30$, a slight difference can be observed even though the actual material parameters remain the same [cf. Fig.65(b)]. Consequently, a precise determination of the thermal conductivity of the investigated material also requires the distance information for these configurations.

The previous studies assume a bulk-like material on top. However, decreasing the thickness of the material [cf. Fig.63(c)] results in completely different ΔT -lines. The thickness influence of the placed material on the ΔT -lines is shown in Fig.66. For decreasing material thickness, the temperature amplitude increases significantly. Furthermore, thin materials on top lead to a continuously changing slope over the whole frequency regime.

4.2.2 Thin layers

Decreasing the thickness of the placed materials results in thin covering layers [cf. Fig.67].

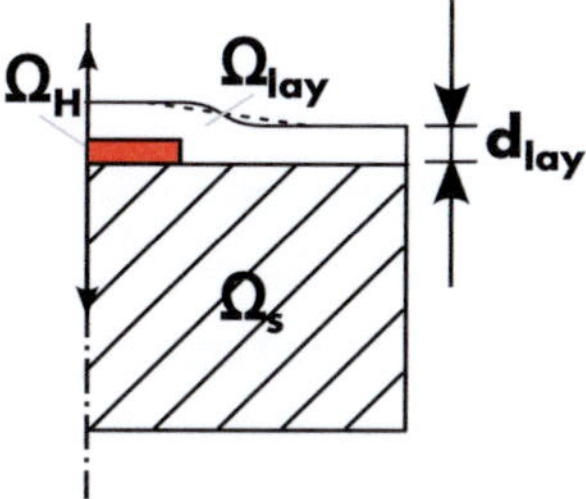

Figure 67: Considered system for the Finite Element simulations for a thin covering layer on top. The dash line indicates contour shape variations of the layer at the outer edge of the heater.

For bulk-like materials on top, the thermal conductivity is clearly represented in the slope of the ΔT -line [cf. Sec.4.2.1]. For such a system, a high contrast in thermal conductivities from the material on top to the substrate would be ideal to determine the thermal conductivity of the investigated material. However, in case of thin layers, the present thermal mass of the placed material is rather small in comparison to the substrate material. Applying a thin poorly-conductive layer material on top of a highly-conductive substrate material results in almost no change for the obtained ΔT -line. This is due to the small present thermal mass and the relatively high thermal resistance of the layer material in comparison to the substrate material. Here, a tremendous amount of heat is conducted through the substrate. Hence, the influence of the thin layer on the temperature amplitude is rather small. **To increase the absolute temperature amplitude and thus slight changes in the ΔT -line due to the layer, one must choose a poorly-conductive substrate material.**

In the following, the influence of the layer's contour shape, the sensitivity for moderately- and poorly-conductive layers and the influence of anisotropic thermal conductivity of the layer is examined on the temperature amplitude. First, the contour shape is considered. To maximize an eventual influence, a thin layer is applied. Fig.68 shows the ΔT -lines for three different shape variations at the outer edge of the heater [cf. Fig.67]. Apparently, the ΔT -lines are coincident and thus, the contour shape can be neglected. Second, the sensitivity for an moderately- (STO) and poorly-conductive (PCMO) layer on top of the YSZ substrate is considered in Fig.69(a) and 69(b), respectively. While a difference occurs for the STO layer, there is almost no difference in the ΔT -lines for the PCMO layer, even though the layer thickness is varied over a wide range. Third, an anisotropy variation [cf. Fig.70] bares, that even tremendous variations of the cross-plane thermal conductivity are absolutely not detectable. On the contrary, a highly-conductive in-plane thermal conductivity could be determined for both layer configurations. However, since thermoelectric materials are expected to be poorly-conductive, these two cases are rather less important for the anisotropy determination. According to the studies for bulk-like materials, the question arises whether additional material above the layer could emphasize the thermal conductivity of the investigated layer. This issue is examined in the next section.

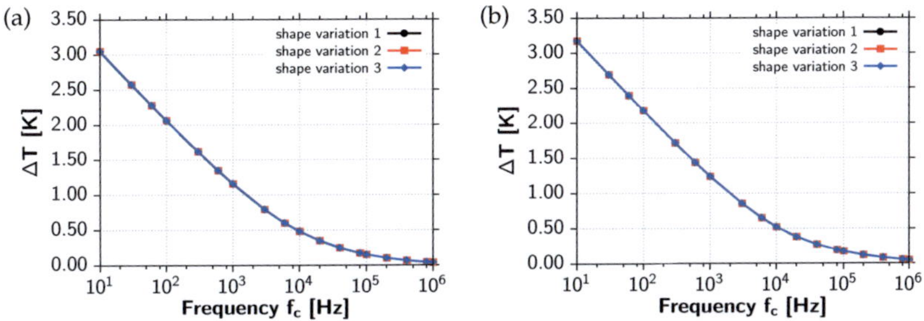

Figure 68: Shape variation for a thin covering (a) STO and (b) PCMO layer on top of a YSZ substrate. The system properties are: $P/L = 5$ W/m, Pt heater, $a = 100$ nm, $2b = 10$ µm, $d_{lay} = 100$ nm, $d_s = 1$ mm, htc $= 5 \cdot 10^7$ W/m^{2}K.

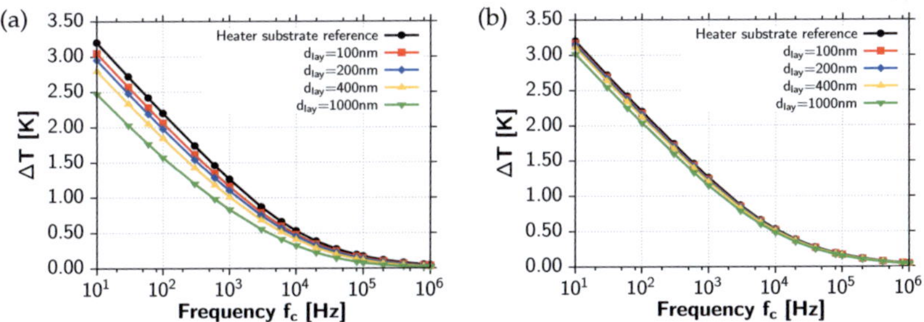

Figure 69: Sensitivity for thin covering (a) STO and (b) PCMO layers on top of a YSZ substrate. The system properties are: $P/L = 5$ W/m, Pt heater, $a = 100$ nm, $2b = 10$ µm, $d_s = 1$ mm, htc $= 5 \cdot 10^7$ W/m^{2}K.

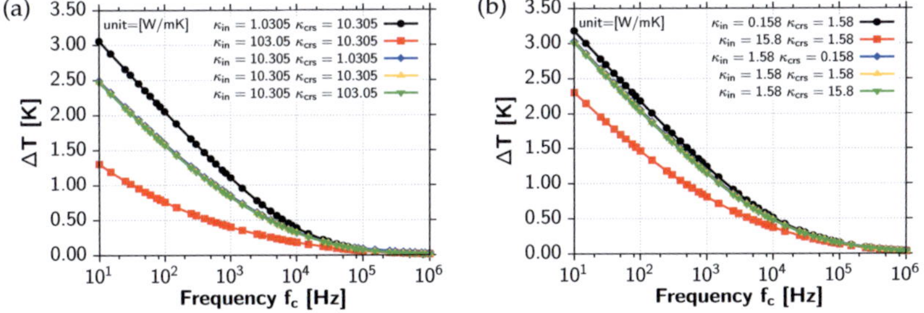

Figure 70: Anisotropic thermal conductivity variation for a thin covering (a) STO and (b) PCMO layer on top of a YSZ substrate. The system properties are: $P/L = 5$ W/m, Pt heater, $a = 100$ nm, $2b = 10$ µm, $d_{lay} = 1000$ nm, $d_s = 1$ mm, htc $= 5 \cdot 10^7$ W/m^{2}K.

4.2.3 Application of heat sink

In this section, the application of a heat sink on top is examined. Basically, the heat sink concept is to enhance the influence of the thin layer on the temperature amplitude in different frequency regimes.

As shown in Fig.64(a), a highly-conductive bulk-like material above the heater substrate platform decreases the temperature amplitude significantly. Thus, a poorly-conductive layer material acts as a thermal resistor between the substrate and the bulk-like material above.

Besides thermal properties, geometrical variations of the heat sink are also essential for the general system behavior. In this work, three different configurations are considered [cf. Fig.71].

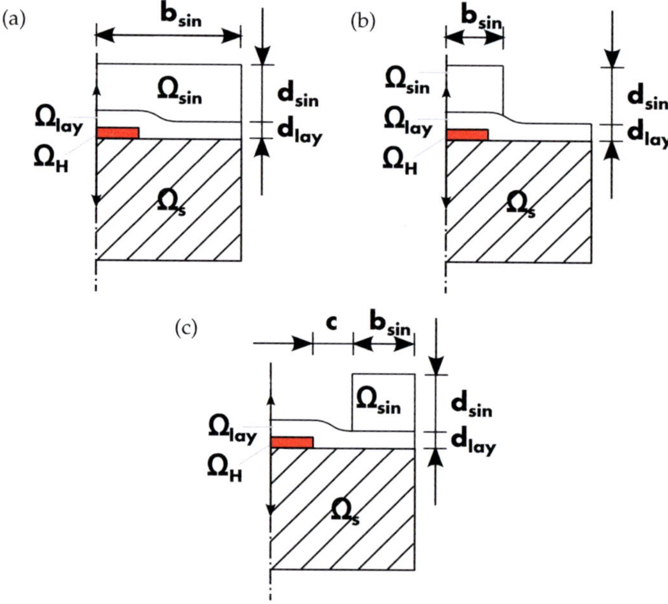

Figure 71: Considered systems with heat sink on top: (a) Full covering heat sink; (b) narrow heat sink directly above the heater; (c) heat sink applied next to the heater.

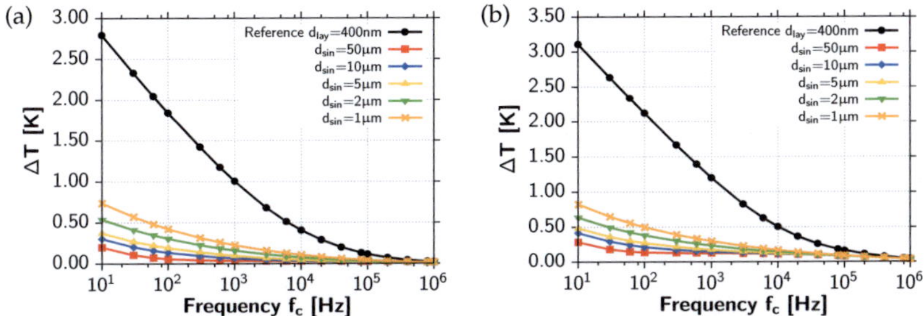

Figure 72: Height variation for a full covering heat sink. A thin covering (a) STO and (b) PCMO layer is placed between the sink and the substrate. The system properties are: P/L = 5 W/m, Pt heater, a = 100 nm, 2b = 10 μm, d_{lay} = 400 nm, Cu heat sink, d_s = 1 mm, htc = $5 \cdot 10^7$ W/m^{2}K.

The intention of different geometries is to increase the influence of either the cross- or in-plane thermal conductivity on the temperature amplitude in different frequency regimes. First, a full covering heat sink is examined for the anisotropic thermal conductivity determination. Here, a 'full covering' heat sink does not equal the substrate width [Fig.13], because the voltage and current pads must not be covered by the material on top of the heater structure. In the following systems, a full covering heat sink has a width of $2b_{sin}$ = 500 μm. Also the height of the heat sink is limited. Due to fabrication steps and the general feasibility, thin copper (Cu) foils are desired as heat sink material. Therefore, we limit the height of the heat sink to d_{sin} = 50 μm.

Apparently, applying a heat sink on top changes the system behavior for an STO [cf. Fig.72(a)] and PCMO [cf. Fig.72(b)] layer significantly. Since the heat sink is placed on top of the structure, a misalignment in positioning could occur. The influence on the ΔT -line is shown in Fig.73.

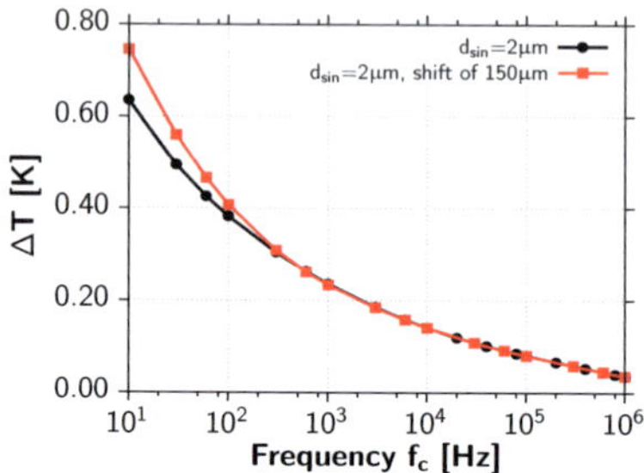

Figure 73: Comparison of a 500 µm broad centrosymmetric aligned heat sink and heat sink with a 150 µm side shift. The system properties are: P/L = 5 W/m, Pt heater, a = 100 nm, 2b = 10 µm, PCMO layer, d_{lay} = 400 nm, Cu heat sink, d_s = 1 mm, htc = 5 · 10^7 W/m²K.

In relation to the tremendous shift of 150 µm to the side, the observed difference is rather small. **In case a slight misalignment is detected after positioning, higher measuring frequencies can counteract the shift.**

Obviously, the heat sink changes the ∆T -lines significantly. However, since the aim is to determine the anisotropic thermal conductivity of the layer, the question arises whether the sensitivity for the layer is enhanced. Therefore, the anisotropic variation is shown in Fig.74. Here, a full covering heat sink with a height of 2 µm is applied.

In comparison to a pure applied layer on top [cf. Fig.70], the influence of a poor cross-plane thermal conductivity is especially enhanced by this heat sink configuration. On the contrary, this system is completely insensitive for in-plane variations. **Consequently, this heat sink configuration is exclusively suitable for poorly-conductive cross-plane determinations.**

If the heat sink width is decreased, the system behaviour changes towards the behavior of thin films on top. However, the heat sink material is still present in the observed ∆T -line.

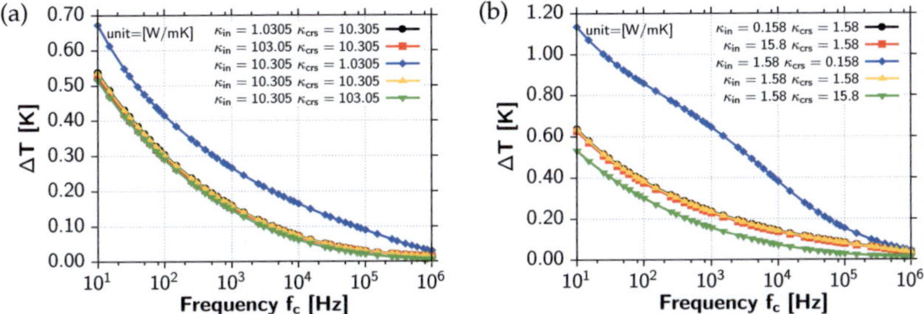

Figure 74: Full covering heat sink with anisotropic variation of the layer. A thin covering (a) STO and (b) PCMO layer is placed between the sink and the substrate. The system properties are: P/L = 5 W/m, Pt heater, a = 100 nm, 2b = 10 μm, d_{lay} = 400 nm, Cu heat sink, d_{sin} = 2 μm, d_s = 1 mm, htc = $5 \cdot 10^7$ W/m²K.

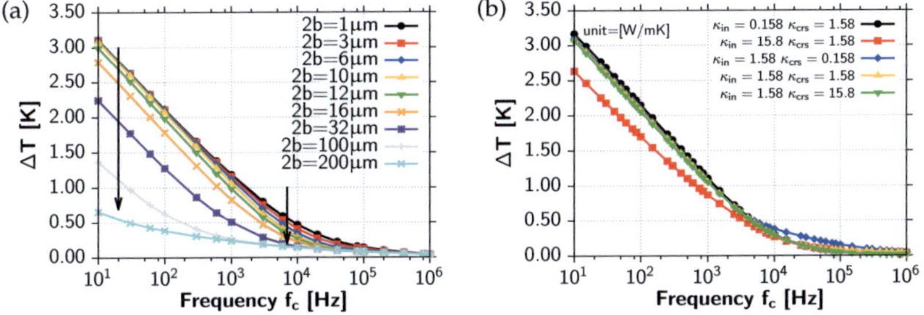

Figure 75: Influence of (a) different heat sink width on top of a PCMO layer and (b) anisotropic variation of the PCMO layer for a sink width of $2b_{sin}$ = 10 μm. The system properties are: P/L = 5 W/m, Pt heater, a = 100 nm, 2b = 10 μm, d_{lay} = 400 nm, Cu heat sink, d_{sin} = 2 μm, d_s = 1 mm, htc = $5 \cdot 10^7$ W/m²K.

In Fig.75(a), the heat sink width is varied tremendously. The dashed lines indicate two different impacts on the ΔT -line. First, a heat sink width up to the heater width, decreases the ΔT -lines just in the high frequency regime while the ΔT -lines are almost coincident in the low frequency regime. Second, a heat sink width bigger than the heater width also affects the ΔT -lines significantly in the low frequency regime.

Based on this behavior, it is examined in Fig.75(b), whether a narrow heat sink can increase the influence of either the cross- or in-plane thermal conductivity at different frequencies. In comparison to a full covering heat sink [cf. Fig.74(b)], a relatively highly-conductive in-plane heat spread is still detectable. However, a poorly-conductive cross-plane thermal conductivity changes the ΔT -lines just in the high frequency regime. **Consequently, these configurations are rather inconvenient to determine the anisotropic thermal conductivity.**
The third conceivable heat sink configuration places the heat sink material next to the heater on top of the layer [cf. Fig.71(c)]. Apparently, this configuration should enhance especially highly-conductive in-plane thermal conductivities. The general system behavior is shown in Fig.76(a) and Fig.76(b) for a small and big distance from heater to heat sink respectively. While a big distance has no profound advantage in comparison to pure thin covering layers on top, a short distance fans out the different anisotropic cases in the low frequency regime. Here, a distance of 1 µm is realized. However, the distance of $c = 10$ µm and the distance of $c = 1$ µm result in profoundly distinct ΔT -lines. Thus, the anisotropic thermal conductivity determination would depend essentially on the contact distance along the heater's length. But since these contact distances are around 0.1% of the heater's length, a slight misalignment could occur. In this case, an identification of the anisotropic thermal conductivity is rather impossible because the ΔT -lines for a contact distance of $c = 1$ µm and $c = 10$ µm are already significantly different.
Furthermore, it has to be mentioned, that the thermal conductivity of thin Cu materials is remarkably thickness dependent [cf. 109]. Also the heat capacity changes with the thickness of the applied heat sink [cf. 167]. However, both thickness dependencies are not considered here.

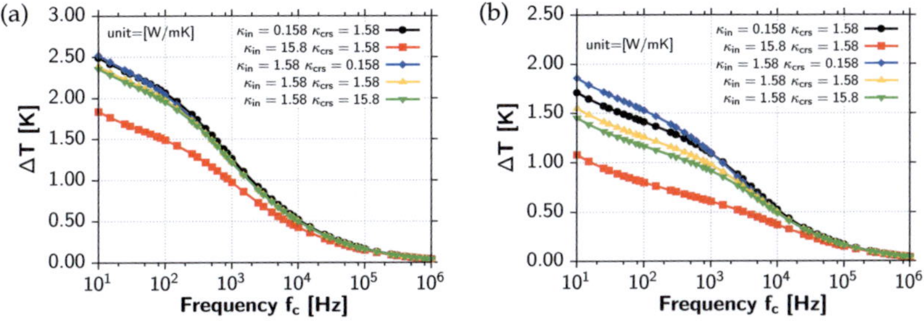

Figure 76: Influence of anisotropic variation for a (a) far placed heat sink with a distance c = 10 μm and a (b) near placed heat sink with a distance c = 1 μm. The heat sink width on each side of the heater is 500 μm − 2b − c. The system properties are: P/L = 5 W/m, Pt heater, a = 100 nm, 2b = 10 μm, d_{lay} = 400 nm, Cu heat sink, d_{sin} = 2 μm, d_s = 1 mm, htc = 5 · 10^7 W/m²K.

4.2.4 Patterned multilayer structures

Theoretically, it is also possible to place nanopatterned materials on top of a heater substrate platform such as in Fig.77.

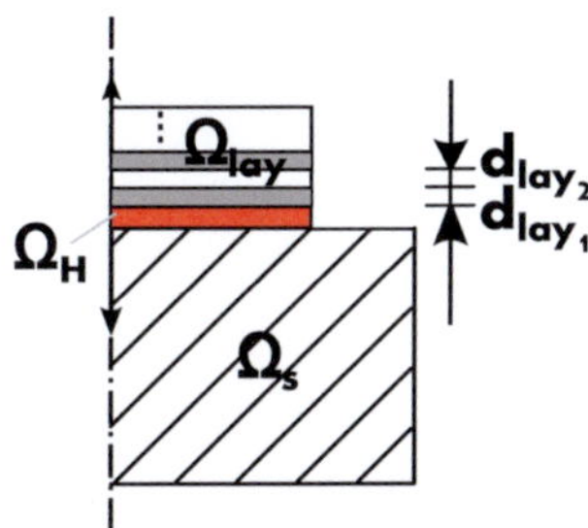

Figure 77: Considered system for nanopatterns on top.

The sensitivity for various patterned PCMO and STO combinations on top of a poorly- and highly-conductive substrate material is shown in Fig.78. The resulting ΔT -lines are almost exclusively affected in the high frequency regime.

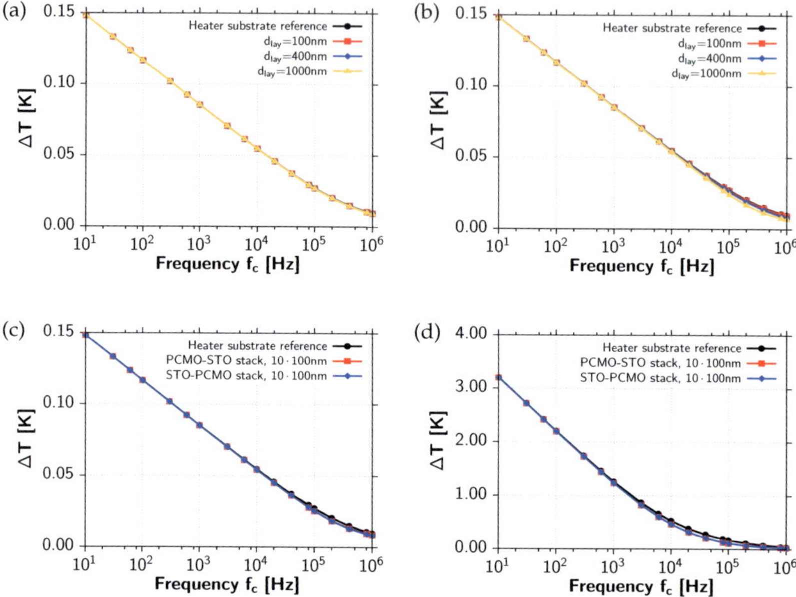

Figure 78: Sensitivity of patterned materials. Influence of placing (a) PCMO, (b) STO material on top of a MgO substrate. Fig.(c) and Fig.(d) show patterned multilayer stacks on a MgO and YSZ substrate respectively. Here, the width of the material equals the heater's width. The system properties are: $P/L = 5\,\text{W/m}$, Pt heater, $a = 100\,\text{nm}$, $2b = 10\,\mu\text{m}$, $d_s = 1\,\text{mm}$, $\text{htc} = 5 \cdot 10^7\,\text{W/m}^2\text{K}$.

For significant changes, enormous stack heights would be required for poorly-conductive thermoelectic materials. Moreover, these ideal systems are rather difficult to manufacture. However, if these structures could be manufactured, relatively high power per length would be required in combination with measurements up to 1 MHz, in order to obtain temperature amplitude variations bigger than the error in the experimental measurement.

4.3 Membrane structures

Every previously investigated geometric structure contained a bulk-like substrate geometry. While most of these geometry configurations are suitable to determine the cross-plane thermal conductivity, not every configuration is adequate to determine the in-plane thermal conductivity. Moreover, every configuration is almost insensitive for poorly-conductive in-plane thermal conductivities. However, since we focus especially on poorly-conductive materials, a general configuration is required to also determine the poorly-conductive in-plane case. Therefore, membrane structures are investigated in this section.

First, two-dimensional membrane systems are considered to develop an understanding of how temperature amplitude lines differ from bulk-like substrates. Here, only thermal simulations are examined. In the beginning, every outer boundary is considered to be adiabatic expect the boundary to the substrate mount. This boundary takes into account a htc. However, since the temperature amplitudes are expected to be significantly higher than for bulk-like substrates with the same applied power per length, radiation losses are also studied. Therefore, we apply the radiation boundary condition of Eq.4.6 at the outer boundaries of the membrane.

Second, three-dimensional membrane systems are examined. Here, we consider every boundary, except the substrate mount to be adiabatic. The substrate mount involves a htc. Moreover, we also consider the mechanical field of Sec.3.1.3, to examine occurring displacements, deflections, and stresses. The thermal and mechanical boundary conditions are given in Sec.3.

4.3.1 Two-dimensional models

Membrane systems with a great length to width ratio are considered as two-dimensional models. For these membrane systems, the in-plane heat spread refers clearly to the perpendicular direction of the symmetry line in Fig.79. The system behavior for different applied power per length is shown in Fig.80 for a membrane system without the application of an additional layer material. The system sketch for this configuration is shown in Fig.79.

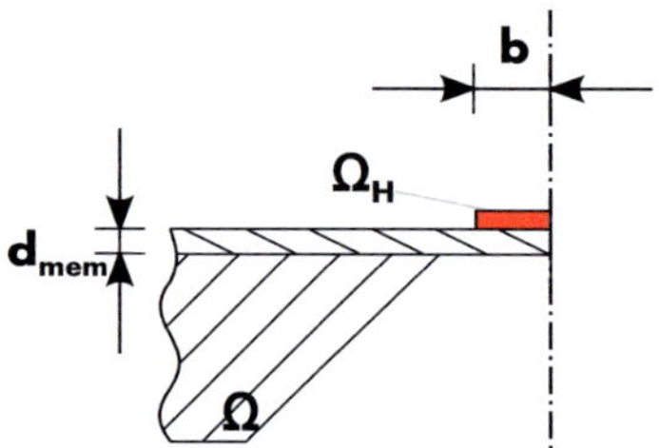

Figure 79: Membrane configuration without the application of an additional layer material.

Since less material is present below the heater, less power per length can be applied to obtain relatively high temperature amplitudes. Furthermore, the inset contains the observed difference between a system taking into account radiation and a system without radiation. Here, the worst case is studied, so that the emissivity is e = 1 and a power per length of P/L = 0.1 W/m is applied. Comparing the temperature amplitude with the difference, an error of less than 1% occurs. Like in Sec.4.1.1.7, this result proves again that radiation errors are negligible for a 3ω-measurement.

In the following, the previous membrane system is considered to point out the mean average temperatures of the heater for a power per length of P/L = 0.01 W/m. Independent of the frequency, the heater's mean average temperature is always around 5 K.

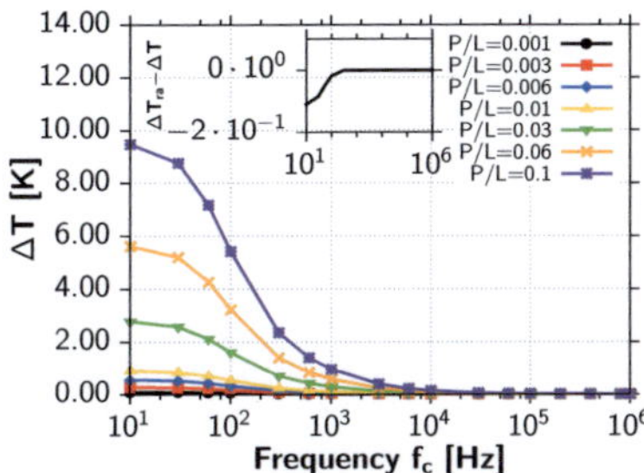

Figure 80: Comparison of different applied power per length P/L. The inset contains the difference for a case with radiation at all outermost boundaries ΔT_{ra} and without radiation ΔT for a P/L = 0.1 W/m. The system properties are: Pt heater, a = 100 nm, 2b = 10 μm, silicon nitride (SIN) membrane, $2b_{mem}$ = 100 μm, d_{mem} = 100 nm, silicon (Si) substrate, d_s = 0.5 mm, htc = $5 \cdot 10^7$ W/m²K.

However, the maximum mean temperature is approximately 10 K for a frequency of f_c = 10 Hz and approximately 5 K for a frequency of f_c = 1 MHz.

To investigate thin layer materials, three distinct configurations exist in general [cf. Fig.81]. The investigated material is either placed between the heater and the membrane, placed on top of the heater membrane system, or placed below the membrane.

The general system behavior for different membrane widths is shown in Fig.82. According to Fig.81, three different layer applications are shown in comparison to a pure membrane system. Basically, a broad membrane implies a higher temperature amplitude. Thus we conclude, that small variations in the anisotropy change the ΔT signal more significantly because the absolute values of ΔT are bigger than for small membranes. In general, this does not depend on the different layer application. Almost no difference occurs between the different systems. In Fig.82, a poorly-conductive layer is applied. The ratio between the thermal conductivity of the layer material to the membrane material is $\kappa_{lay}/\kappa_{mem} \approx 0.66$.

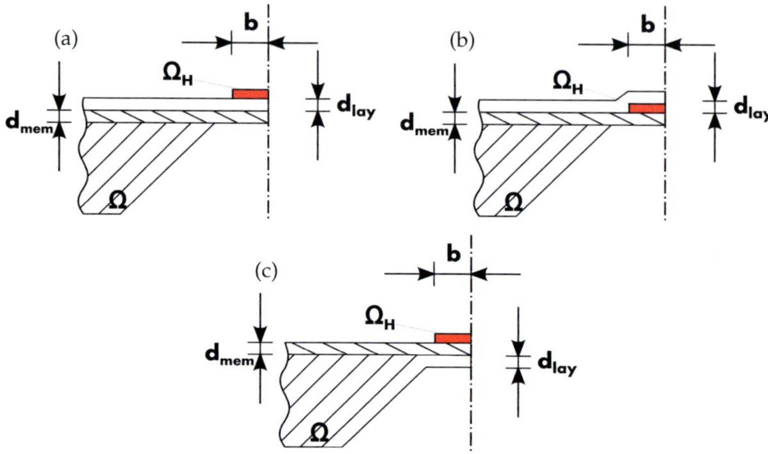

Figure 81: (a) Investigated layer between heater and membrane; (b) layer placed on top of heater membrane structure; (c) layer placed below the heater membrane structure.

A highly-conductive layer material with a bigger $\kappa_{lay}/\kappa_{mem}$ ratio increases the difference between the various layer applications slightly. After examining the general system behavior, different membrane thicknesses are studied for the sensitivity to anisotropic thermal conduction. Since the membrane is only used for the purpose of investigating the layer material, its desired thickness has to be as thin as possible. Furthermore, a non-conductive membrane would be ideal to isolate the thermal properties of the layer material. Because in this case, the temperature amplitude would only depend on the thermal properties of the layer to be investigated. However, the membrane must obviously exhibit thermal properties and a thickness.

The sensitivity for a poorly- and moderately-conductive layer material is examined in Fig.83. First, a comparison of Fig.83(a)83(c)83(e) and Fig.83(b)83(d)83(f) bares, that a thinner membrane increases the temperature amplitude in general for both investigated materials.

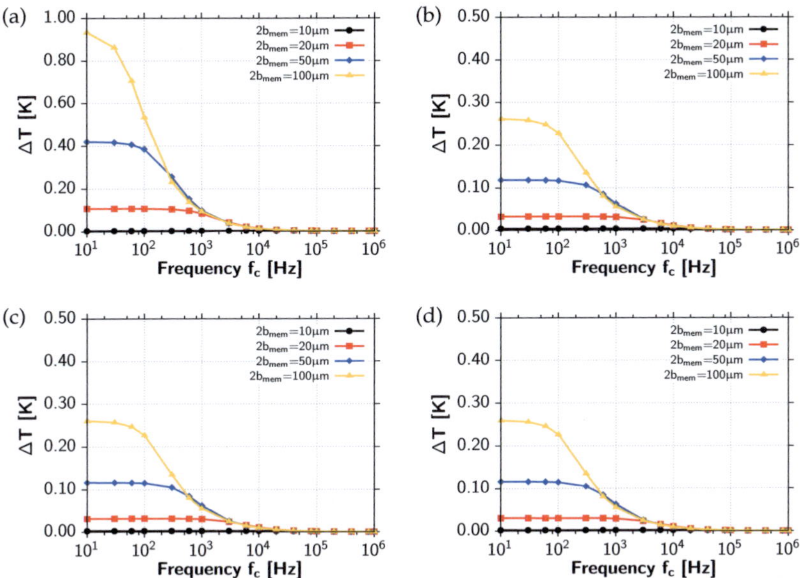

Figure 82: System behavior for two-dimensional membrane systems. (a) Pure membrane on substrate system, (b) layer between the heater and the membrane, (c) layer placed on top of heater and membrane and (d) layer placed below the membrane. The system properties are: P/L = 0.01 W/m, Pt heater, a = 100 nm, 2b = 10 μm, PCMO layer, d_{lay} = 400 nm, SIN membrane, d_{mem} = 100 nm, Si substrate, d_s = 0.5 mm, htc = $5 \cdot 10^7$ W/m^{2}K.

Second, these membrane systems are almost exclusively sensitive to anisotropic in-plane thermal conductivity variations. Cross-plane variations do not affect the temperature amplitudes. Moreover, differences in ΔT occur only in the low frequency regime. Consequently, the dimensions of the membrane limit essentially the sensitivity for the in-plane conduction. **However, we conclude that membrane configurations which are fabricated via micro electro mechanical systems (MEMS), with a bigger membrane width than $2b \geq 100$ μm and a thickness less than $d_{mem} \leq 100$ nm are suitable to determine the in-plane thermal conductivity for both materials PCMO and STO.**

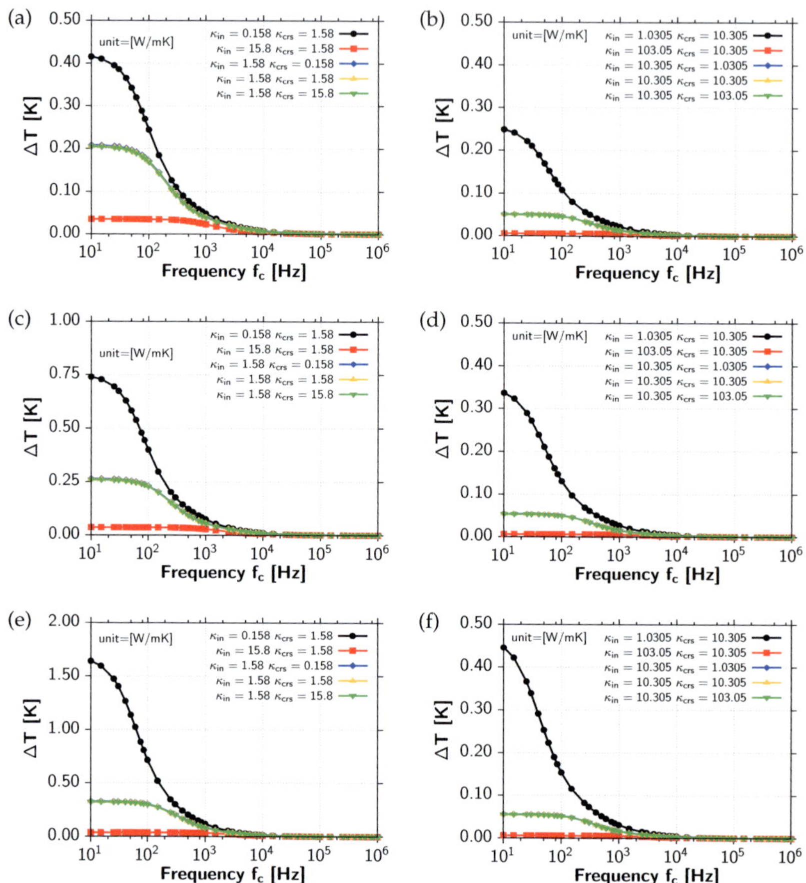

Figure 83: Influence of membrane thickness for anisotropic layer material. Sensitivity for an anisotropic PCMO layer material with a membrane thickness of (a) 200 nm, (c) 100 nm and (e) 30 nm. Sensitivity for an anisotropic STO layer material for a membrane thickness of (b) 200 nm, (d) 100 nm and (f) 30 nm. The system properties are: P/L = 0.01 W/m, Pt heater, a = 100 nm, 2b = 10 μm, PCMO layer, d_{lay} = 400 nm, SIN membrane, $2b_{mem}$ = 100 μm, Si substrate, d_s = 0.5 mm, htc = $5 \cdot 10^7$ W/m²K.

4.3.2 Three-dimensional 6-pad heater structure

In the previous section, it is shown, that two-dimensional membrane structures are suitable to determine the in-plane thermal conductivity. However, these models ignore the heat conduction along the heater's length dimension. But towards the end of the heater's length, thus behind the membrane area, a bulk-like substrate is present. Here, we apply a Si substrate. Si is regarded as a highly-conductive substrate material. As shown in Sec.4.1.2.1, highly-conductive substrates are especially appropriate to determine the cross-plane thermal conductivity. Thus the question arises, whether a heater structure could combine these properties. Hence it is examined in this section, whether a three-dimensional designed geometric structure has the potential to determine the cross- and in-plane thermal conductivity with one geometry configuration. Moreover, we propose in this work a special heater configuration with six contact pads. The investigated layer material is placed between the heater and the membrane [cf. Fig.81(a)]. The general system sketch is shown in Fig.84.[38]

We call this configuration a 6-pad heater structure. The idea of the 6-pad structure is to have two different measuring configurations. The first configuration is the so-called short heater configuration [cf. Fig.84(b)]. The current is applied through the middle pads (shaded in red) and the voltage is measured at the inner pads (shaded in green). We think that in this configuration, the voltage signal and thus the temperature amplitude is dominated by the influence of the in-plane thermal conductivity. The second configuration is the so-called long heater configuration [cf. Fig.84(c)].

---

[38] A technical drawing of the 6-pad structure is given in App.C.

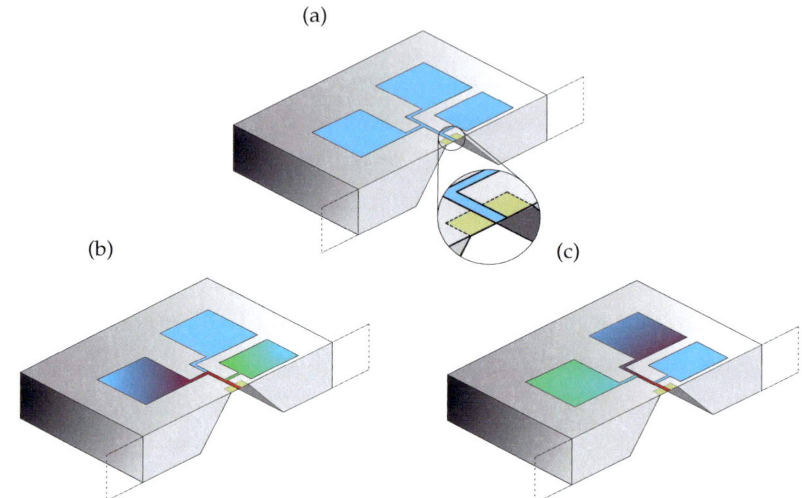

Figure 84: (a) General system sketch for a 6-pad heater structure; (b) short heater configuration; (c) long heater configuration. The applied heating is marked in red, the green color indicates the voltage tapping.

Here, the current is applied through the outer pads and the voltage is measured at the middle pads. In this configuration, the voltage signal and thus the temperature amplitude should also contain significantly the influence of the cross-plane thermal conductivity. However, in both configurations, only four pads are used for each measurement. A main advantage of such a configuration would be the determination of the cross- and in-plane thermal conductivity for one fabricated sample and thus for persisting layer properties. The general system behavior for the 6-pad structure is shown in Fig.85.

First, the STO layer material is considered. In general, both heater configurations [cf. Fig.85(a) and Fig.85(b)] exhibit a similar system behavior in comparison to the two-dimensional model of membrane systems. It turns out that, the sensitivity for in-plane conduction is

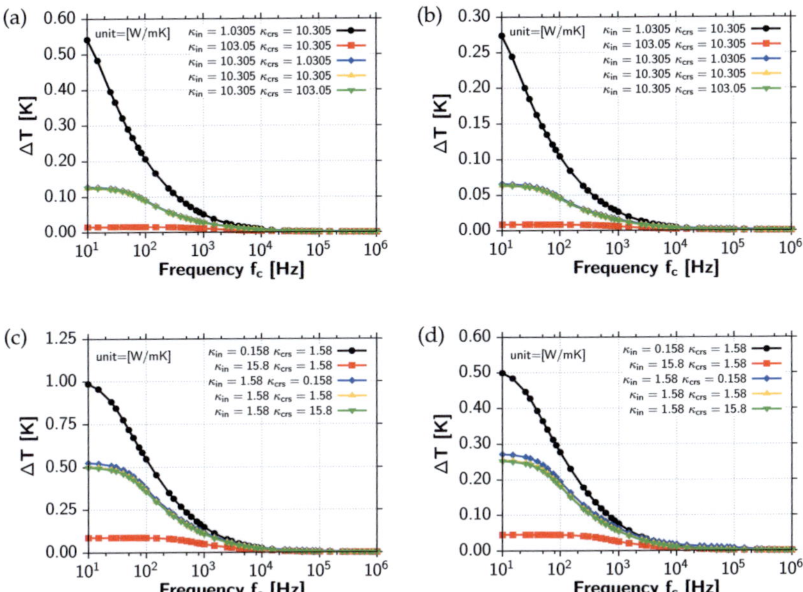

Figure 85: System behavior for a three-dimensional membrane system. Short heater configuration for a (a) STO and (c) PCMO layer. Long heater configuration for a (b) STO and (d) PCMO layer. The system properties are: $P/L = 0.05$ W/m, Pt heater, $a = 100$ nm, $2b = 10$ µm, PCMO layer, $d_{lay} = 500$ nm, SIN membrane, $d_{mem} = 100$ nm, $2b_{mem} = 250$ µm, Si substrate, $d_s = 0.5$ mm, htc $= 5 \cdot 10^7$ W/m^{2}K.

higher for the short heater configuration, than for the long heater configuration. Moreover, a comparison of the long and short heater configuration exhibits, that the temperature amplitude in the long heater configuration is smaller than in the short heater configuration. This is because more thermal mass is present below the long heater. Second, the PCMO layer material is considered. Also here, the ΔT-lines for a short heater configuration are similar to the ΔT-lines of the two-dimensional model of membrane systems.

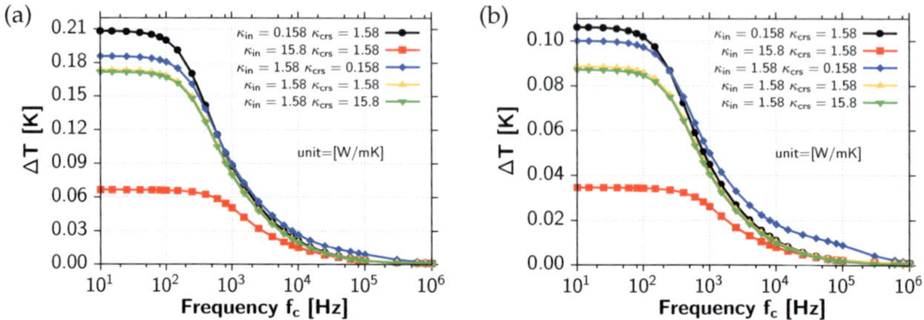

Figure 86: System behavior for a three-dimensional membrane system. The resulting temperature amplitudes are shown in (a) for the short heater configuration and in (b) for the long heater configuration. The system properties are: $P/L = 0.05$ W/m, Pt heater, $a = 100$ nm, $2b = 10$ µm, PCMO layer, $c_{P_{lay}} = 48$ J/kgK, $d_{lay} = 500$ nm, SIN membrane, $\kappa_{mem} = 30.1$ W/mK, $d_{mem} = 100$ nm, $2b_{mem} = 250$ µm, Si substrate, $d_s = 0.5$ mm, htc $= 5 \cdot 10^7$ W/m^{2}K.

However, for both materials the long heater configuration is not suitable to determine also the cross-plane thermal conductivity.

Consequently, the proposed 6-pad heater structure can be applied to determine the in-plane thermal conductivity for both materials STO and PCMO. But, it is not possible to determine the in- and cross-plane thermal conductivity of the STO and PCMO layer simultaneously. However, in this work, we use a poorly-conductive SIN membrane. But in literature, also moderatley-conductive SIN membranes are reported with a thermal conductivity of $\kappa_{mem} = 30.1$ W/mK [46]. Based on such an SIN membrane material, we investigate further, if it would be possible to determine the cross- and in-plane thermal conductivity with the 6-pad structure. Moreover, for this investigation, we apply a PCMO like layer material where the heat capacity is decreased by a factor of ten so that $c_p = 48$ J/kgK. The resulting temperature amplitudes are shown in Fig.86. While the system behavior for the short heater configuration is similar to Fig.85(c), the system behavior

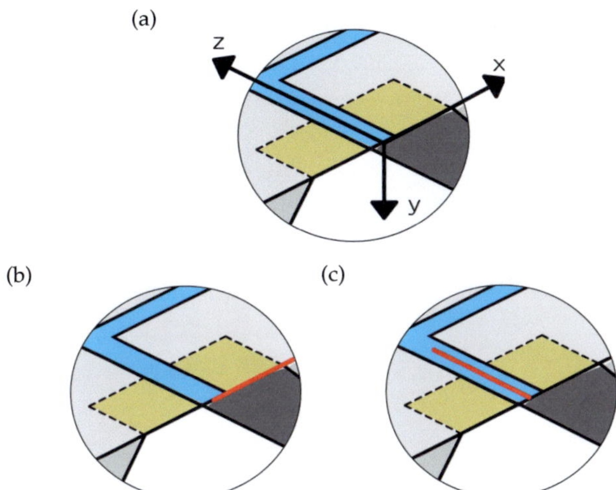

Figure 87: (a) Illustration of the coordinate system, (b) the horizontal cut and (c) the cut along the heater. The point of origin (0,0,0) is defined between the heater and the layer material.

for the long heater configuration exhibits a distinctly different ΔT -line for the extremely poorly-conductive cross-plane case [cf. Fig.86(b)]. Here, the ΔT -line shifts up over the whole frequency regime.

Thus essentially, the cross- and in-plane thermal conductivity of such an extremely poorly-conductive layer material could be determined by the proposed 6-pad heater structure with a moderately-conductive SIN membrane.[39] Here, the same power per length is applied than for the poorly-conductive membrane so that Fig.86 is comparable to Fig.85. However, since the membrane's thermal conductivity is higher here, the temperature amplitudes in Fig.86 are rather small. By applying a power per length around P/L = 0.1 W/m, temperature amplitudes up to one Kelvin can be achieved, but essentially, the system behavior remains the same.

---

[39]If the heat capacity of the layer is c_p = 480 J/kgK, the blue ΔT -line in Fig.86 would decrease slightly in the high frequency regime, but the ΔT -line still differs significantly from the other anisotropic cases.

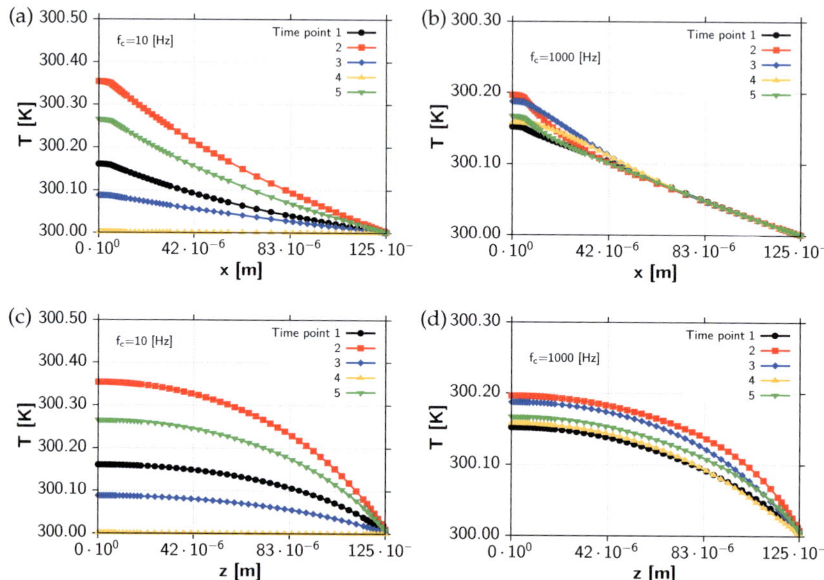

Figure 88: Temperature distribution along a (a),(b) horizontal and (c),(d) length cut. The cutting lines are located in the center of the layer. The system properties are: P/L = 0.05 W/m, Pt heater, 2b = 10 μm, a = 100 nm, d_{lay} = 500 nm, STO layer material, $2b_{mem}$ = 250 μm, SIN membrane, Si substrate, d_s = 0.5 mm and htc = $5 \cdot 10^7$ W/m²K.

In the following, the spatial temperature distribution is examined for the STO layer material to understand the system behavior. According to Sec.4.1.1.4, five different time points are considered. In contrast to two-dimensional model systems, one cut is regarded from the center point of the heater towards the perpendicular side of the membrane. The other cut is along the heater towards the support of the substrate. Both cuts are situated in the center of the layer material. Fig.87 illustrates these two cuts. The temperature distribution is shown in Fig.88 for these two cuts.

Here, Fig.88(a) and 88(c) show clearly decreasing temperature trends for each time point without an overlap.   On the contrary, the temperature lines in Fig.88(b) and 88(d) cross each other at a certain spatial length. For these systems, the spatial temperature distribution does not decrease as fast as for bulk-like materials. The reason is that, less material is present below the heat source.

The cut along the heater's length dimension exhibits a completely different temperature distribution. Independent from the frequency, the spatial temperature amplitude decreases significantly towards the substrate support. This behavior indicates a high sensitivity for the in-plane thermal conduction and a relatively low sensitivity for the cross-plane conduction above the substrate support. This is because, the temperature amplitudes are in general tremendously smaller than in the area of the membrane. The overall temperature distribution is shown in Fig.93 for the short heater configuration. Since a MEMS applies a membrane with a big width to thickness ratio, the deflection and occurring stresses within a measurement are investigated in the following. According to results, the maximum deflection occurs at the center of the membrane. Therefore, we observe the displacement of a single point located between the heater and the membrane. Fig.89 illustrates the considered point for a short heater configuration.

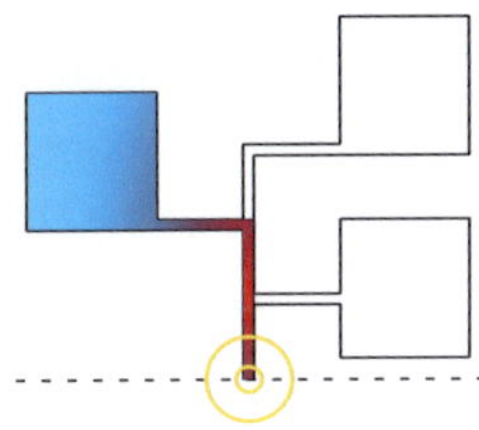

Figure 89: Illustration of the considered point of deflection at (0,0,0) for the short heater configuration.

133

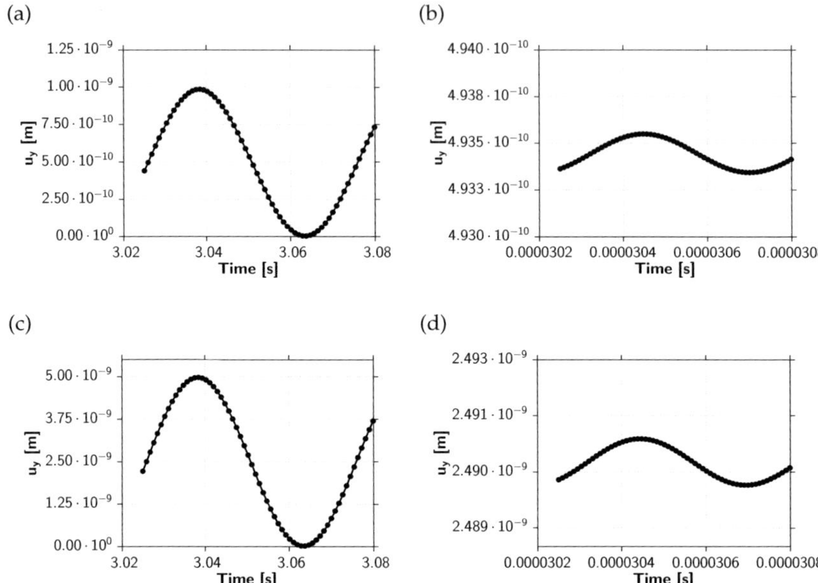

Figure 90: Deflection of the center point for a three-dimensional membrane system with the short heater configuration. Deflection for a STO and PCMO layer at (a)(c) 10 Hz and (b)(d) 1 MHz respectively. The system properties are: $P/L = 0.05$ W/m, Pt heater, $a = 100$ nm, $2b = 10$ µm, $d_{lay} = 500$ nm, SIN membrane, $d_{mem} = 100$ nm, $2b_{mem} = 250$ µm, Si substrate, $d_s = 0.5$ mm, htc $= 5 \cdot 10^7$ W/m^{2}K.

The deflections of the center point are shown in Fig.90. The deflection necessarily follows the temperature behavior in general for both investigated materials. While the deflection returns almost to zero for a frequency of $f_c = 10$ Hz, the displacement remains at a higher deflection for a frequency of $f_c = 1$ MHz. However, for both investigated materials, the deflections of the center point are certainly less than 10 nm. Moreover, the maximum displacements of the other spatial dimensions are even smaller [cf. Fig.93].

Considering the mechanical field also allows to examine occurring stresses besides displacement observations. The occurring stresses are calculated by [8, p. 54]

$$\sigma_x = 2Gu_{x_x} + \frac{2G\nu}{(1-2\nu)}\left(u_{x_x} + u_{y_y} + u_{z_z}\right) - \frac{E_\sigma}{1-2\nu}\beta(T - T_0), \quad (4.13)$$

$$\sigma_y = 2Gu_{y_y} + \frac{2G\nu}{(1-2\nu)}\left(u_{x_x} + u_{y_y} + u_{z_z}\right) - \frac{E_\sigma}{1-2\nu}\beta(T - T_0), \quad (4.14)$$

$$\sigma_z = 2Gu_{z_z} + \frac{2G\nu}{(1-2\nu)}\left(u_{x_x} + u_{y_y} + u_{z_z}\right) - \frac{E_\sigma}{1-2\nu}\beta(T - T_0), \quad (4.15)$$

and the shear stresses are

$$\tau_{xy} = G \cdot \left(u_{x_y} + u_{y_x}\right), \quad (4.16)$$

$$\tau_{xz} = G \cdot \left(u_{x_z} + u_{z_x}\right), \quad (4.17)$$

$$\tau_{yz} = G \cdot \left(u_{y_z} + u_{z_y}\right). \quad (4.18)$$

In this work, the stresses are discussed without the evaluation of a stress hypothesis. In the following, three different cuts are considered. Fig.91 illustrates these three cuts. Every cut originates in the center at $z = 0$ m. The occurring stresses along these different cuts are shown in Fig.92.

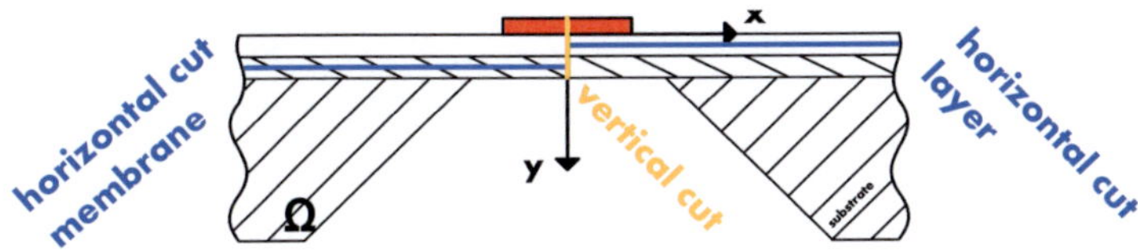

Figure 91: Illustration of the different cuts for the stress evaluation. The horizontal cuts are shown in blue color in the middle of the layer and the membrane. The vertical cut through the heater, the layer and the membrane is shown in yellow color.

(a)

(b)

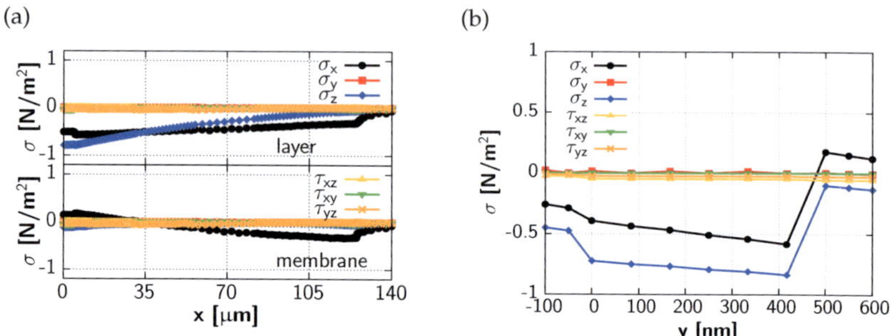

Figure 92: Occurring stress for a cut (a) horizontal from the center point in the middle of the layer and in the middle of the membrane towards the membrane's side and (b) a vertical cut over the height at the center point. The system properties are: P/L = 0.05 W/m, Pt heater, a = 100 nm, 2b = 10 μm, STO layer, d_{lay} = 500 nm, SIN membrane, d_{mem} = 100 μm, $2b_{mem}$ = 250 μm, Si substrate, d_s = 0.5 mm, htc = $5 \cdot 10^7$ W/m²K.

While Fig.92(a) shows the occurring spatial stresses along a horizontal cut line in the middle of the layer and the middle of the membrane, Fig.92(b) shows the occurring spatial stresses over the thickness of the heater layer and membrane package. First, the horizontal cut is considered. The first inset of Fig.92(a) shows the stresses in the middle of the layer. All stresses are almost zero, except the stresses σ_x and σ_z. The layer is under pressure until the substrate mount. On the contrary, the second inset of Fig.92(a) shows that the membrane is under tension for approximately a quarter of the total membrane width. All magnitudes decrease significantly towards the substrate support. Second, the vertical cut is regarded. Fig.92(b) shows the stresses along this cut. All stresses are almost zero, except the stresses σ_x and σ_z. Here, these two stresses differ slightly from each material to the neighboring material. While the heater and the layer are under pressure, the membrane exhibits almost no stress. **However, in general all occurring stress magnitudes are extremely small and we don't expect any material to fail with this small applied power per length.**

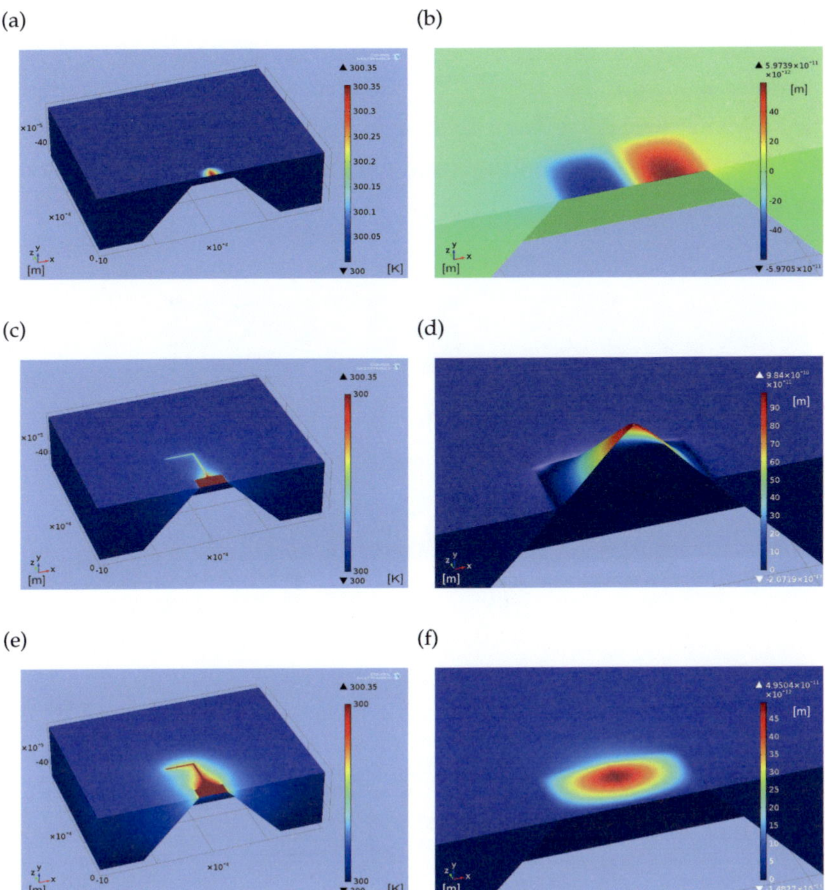

Figure 93: The maximum temperature distribution for the short heater configuration on top of an STO layer is shown in Fig.(a). A different scaling is shown in Fig.(c) and (e) for this temperature distribution. Fig.(b),(d) and (f) show the maximum spatial displacements for the x-, z- and y-direction respectively. The system properties are: P/L = 0.05 W/m, Pt heater, a = 100 nm, 2b = 10 μm, STO layer, d_{lay} = 500 nm, SIN membrane, d_{mem} = 100 nm, $2b_{mem}$ = 250 μm, Si substrate, d_s = 0.5 mm, htc = $5 \cdot 10^7$ W/m^{2}K.

4.4 Pillar and pad structure

Contrary to a 3ω-measurement, the pillar and pad experiments apply the PCMO material itself as a heating material. The three-dimensional structure, the system definitions and the pillar's and pad's dimensions are shown in Fig.94.

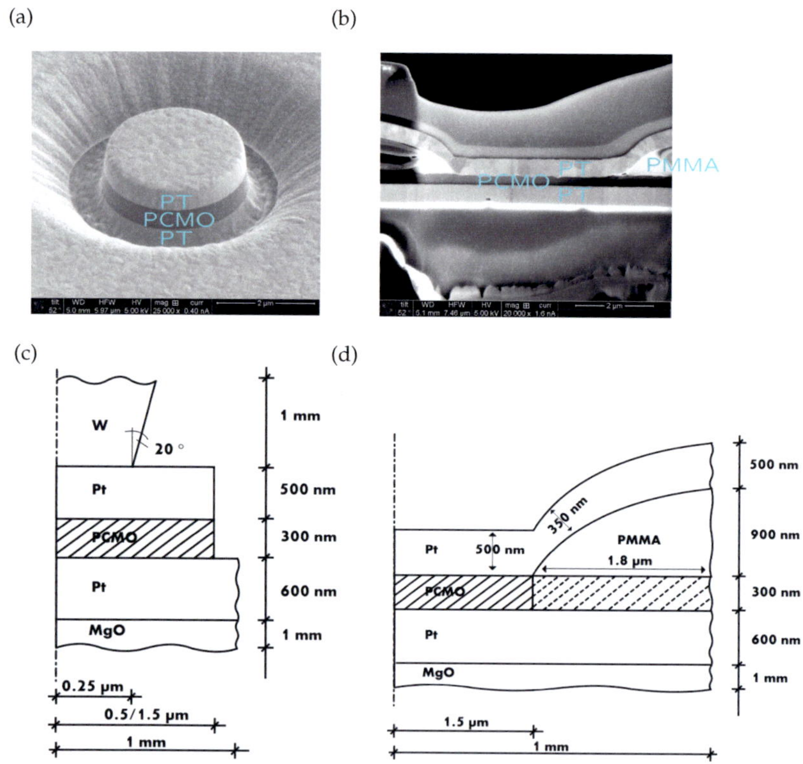

Figure 94: Scanning electron microscopy (SEM) images of the (a) three-dimensional pillar structure and (b) the PCMO pad structure with an electrically insulating PMMA layer. The SEM images were taken by our project partners at the University of Göttingen. The corresponding system definitions and dimensions are given in (c) for the pillar and in (d) for the pad structure.

In the beginning, the PCMO pillar is considered [cf. Fig.94(c)]. Here, we distinguish between two different configurations. First, a pillar where only a small Pt electrode is placed on top of the PCMO material. Second, a pillar where a tungsten (W) needle (Kleindiek) is in thermal contact with the Pt top electrode.[40]

Afterwards, the PCMO pad device is considered [cf. Fig.94(d)]. Here, a PCMO layer is placed between a Pt top and bottom electrode. This sandwiched structure encloses an electrically insulating Polymethyl Methacrylate (PMMA) layer ($C_5H_8O_2$) to realize a pillar like PCMO structure.[41] Although, the Pt materials above and below the PCMO material exist to induce the current, both the pillar and pad investigations, apply Joule heating only in the area of the PCMO material, because the electric resistivity of Pt is negligible in comparison to PCMO.

In contrast to the previous investigations for 3ω-measurements, the pillars exhibit a rotation symmetry which we use for the simulations to reduce computing time. The respective governing equations are given in Sec.3.1.1. Here, the power input is not regarded as a power per length and not applied with an angular frequency. According to Sec.3.1.1, the heat source term of Eq.3.16 writes now to

$$p = \varrho_0(T) \cdot j(t)^2 . \qquad (4.19)$$

Here, $\varrho_0(T)$ is the temperature dependent specific electric resistivity and $j(t)$ is the current density. Details about the application of the time dependent current density are given in Sec.3.2.

---

[40]Find App.B for further information about W.
[41]Find App.B for further information about PMMA.

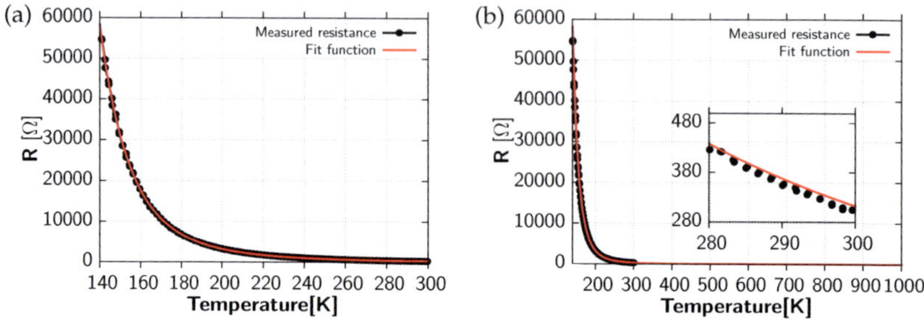

Figure 95: (a)(b) Experimentally measured data points of the electric resistance and corresponding fit function for the PCMO pillar of Fig.94(a) with a radius of $r_p = 3$ μm and a height of $d_{PCMO} = 300$ nm.

The specific electric resistivity is calculated by

$$\varrho_0(T) = \frac{R(T) \cdot A_p}{d_{PCMO}}.$$ (4.20)

Here, $A_p = \pi \cdot r_p^2$ is the circular surface area of the pillar with the radius r_p and $R(T)$ is the electric resistance behavior, which was experimentally measured. Moreover, the electric resistance behavior is strongly non-linear with temperature. The temperature dependent resistivity is shown in Fig.95. Experimental data points are just existent until 300 K.

However, since current densities are applied up to $j = 1 \cdot 10^{10}$ A/m^2, we expect a tremendous temperature increase. In this work, a fit function is used. Thus, the known experimental data is represented and the electric resistance behavior is further extrapolated. The fit function decreases strictly monotonically up to several thousand Kelvin, but always remains positive. The fit function writes to

$$R(T) = 77.86/\sqrt{T} \cdot e^{1272.26/T} \quad [\Omega].$$ (4.21)

Although we do not know the melting point of PCMO, we know that it must be over 1300 K since its thermoelectric properties are characterized up to this temperature [cf. 43]. According to Sec.3.1.1, we assume that no heat flows through the axis of symmetry Γ_1 and the outer boundaries Γ_2 and Γ_3. In contrast to the 3ω-simulations, the substrate mount is considered isothermal so that

$$\Gamma_4 : \quad T = T_0 . \tag{4.22}$$

In the following, three different initial and ambient temperatures are studied

$$\Omega : \quad T_{init} = T_0 = 300, 230, 160 \text{ K} . \tag{4.23}$$

Since the heat capacity of PCMO changes especially for low temperatures [cf. 102], a linear function is used in this section for the heat capacity of PCMO. This funtion writes to

$$c_p = \begin{cases} 0 \leq T \leq 230 & : 2.087 \cdot T \quad [\text{J/kgK}] \\ 230 < T < \infty & : 480 \qquad\quad [\text{J/kgK}] . \end{cases} \tag{4.24}$$

The resulting time dependent temperature behavior for the pillar without the Kleindiek contact is shown in Fig.96. Here the temperature represents the mean temperature in the PCMO material, determined by Eq.3.17.

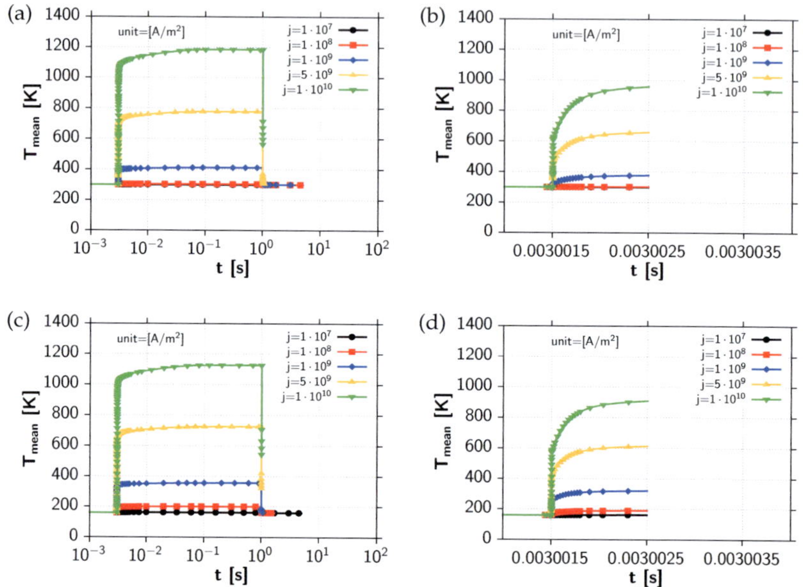

Figure 96: The pillar's time dependent mean temperature behavior for an initial temperature of 300 K for (a) $dur_p = 1$ s and (b) $dur_p = 1$ µs and for an initial temperature of 160 K for (c) $dur_p = 1$ s and (d) $dur_p = 1$ µs.

Fig.96 is discussed in the following. First, independent of the initial temperature and pulse length, a current density of $j = 1 \cdot 10^{10}$ A/m^2 results in the maximum temperature increase. Second, already current densities which are applied for 1 µs result in remarkably high temperatures in comparison to current densities which are applied for 1 s. Thus, the major part of heating takes place in the first micro second. Third, for every initial temperature and simulated current density, the temperature returns back to initial condition within the pulse repetition rate f_p, although temperatures up to 1200 K occur. This fact is shown in Fig.97 for a reference current density of $j = 1 \cdot 10^{10}$ A/m^2.

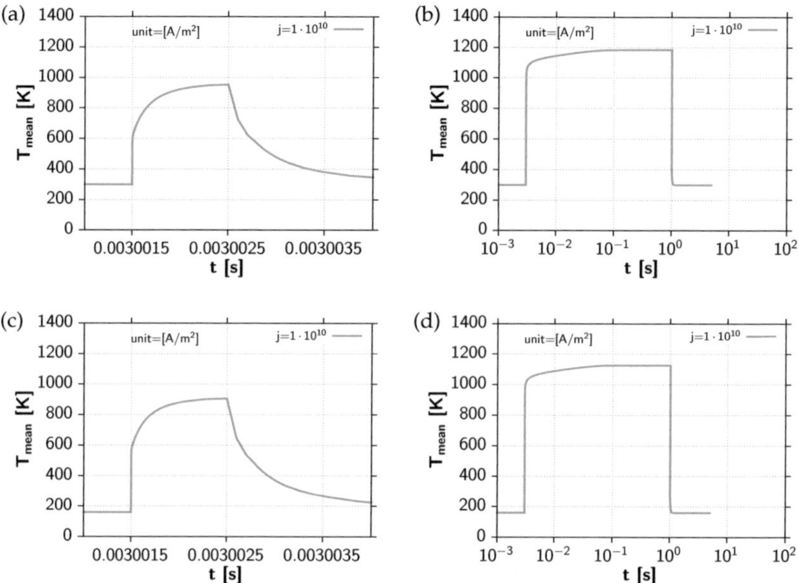

Figure 97: The pillar's time dependent mean temperature behavior for an initial temperature of 300 K for (a) $\mathrm{dur_p} = 1$ µs and (b) $\mathrm{dur_p} = 1$ s and for an initial temperature of 160 K for (c) $\mathrm{dur_p} = 1$ µs and (d) $\mathrm{dur_p} = 1$ s.

In the following we consider only this reference current density. For an initial temperature of $T_{\mathrm{init}} = 300$ K, both pulse lengths indicate a strictly monotonic increasing temperature behavior until the pulse ends [Fig.97(a) and 97(b)]. Switching off the current results in a strictly monotonic decreasing temperature behavior. In fact, the major temperature development occurs within the first $100 - 500$ ns.

In principal, the pillar's mean temperature increases in the same way for an initial temperature of $T_{\mathrm{init}} = 160$ K [Fig.97(c) and 97(d)]. However, although the initial electric resistance is tremendously higher for an initial temperature of $T_{\mathrm{init}} = 160$ K in comparison to an initial temperature of $T_{\mathrm{init}} = 300$ K, the resulting maximum mean temperatures are similar.

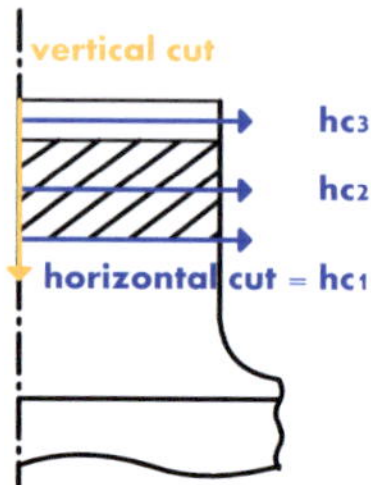

Figure 98: Definition of a vertical cut (yellow) and three horizontal cuts (blue) within the pillar configuration.

To understand the different temperature developments, vertical and horizontal cuts through the pillars are taken into consideration in the following. The cuts are defined in Fig.98. First, the pillar without the Kleindiek contact is considered and the horizontal cut hc1 and the vertical cut are examined. Subsequently, the pad device is examined with two additional horizontal cuts, hc2 and hc3.

Fig.99 shows the temperature profiles for an initial temperature of T_{init} = 300 K at several time steps within the heating pulse. Since basically, both pulse durations exhibit the same temperature development along the vertical cut and the horizontal cut hc1 for the first 1 µs, we consider only a pulse duration of 1 s in the following. However, higher temperatures occur for a pulse duration of 1 s.

The time intervals for a vertical cut indicate a rapid temperature increase in the center of the PCMO layer [cf. Fig.99(a)]. Since it requires some time for heat to diffuse into the top Pt electrode, a time lag of the respective local maximum temperature is visible. Despite this lag, the temperature in the top electrode adapts to the temperature of the PCMO layer before the pulse is completed. In case of the free standing pillar, the maximum temperature occurs at the top of the PCMO layer. The temperature below the PCMO layer increases significantly slower since heat is conducted into the substrate.

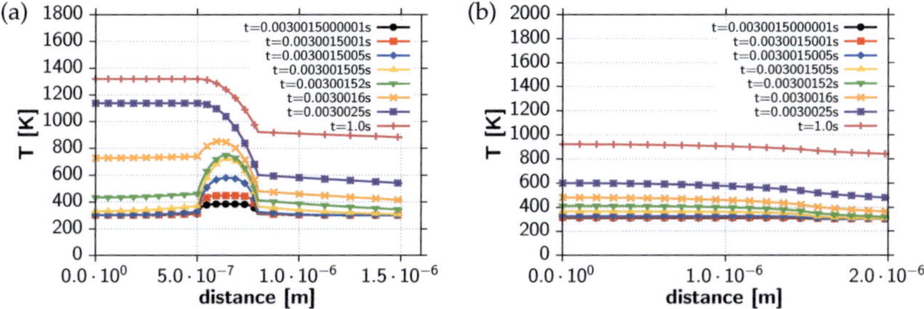

Figure 99: Temperature profiles for the pillar of Fig.94(c) without the Kleindiek contact and an initial temperature of T_{init} = 300 K and a pulse duration of dur_p = 1 s. The vertical cut of 98 is shown in (a) and the horizontal cut hc1 is shown in (b). The temperature profiles are simulated for the reference current density of $j = 1 \cdot 10^{10}$ A/m^2.

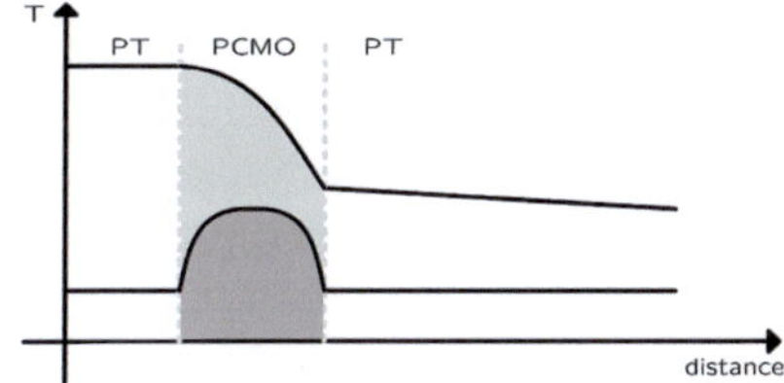

Figure 100: Indicated area below the temperature profile to illustrate the average temperature along the vertical cut for comparison to the determined integral temperature in the PCMO layer.

Moreover, the horizontal cut indicates that the maximum temperature is always at the axis of symmetry and decreases towards the outer side of the pillar [cf. Fig.99(b)]. Due to this fact, the determined mean temperature of the PCMO layer can be correlated to the average temperature along the vertical cut because the temperature gradient is almost zero, so that $\partial T/\partial r \approx 0$. This circumstance is illustrated in Fig.100. Regarding this circumstance, several time points in Fig.99(a) illustrate the strictly monotonic temperature increase of Fig.97(a) and 97(b). Thus, the mean PCMO temperature must increase until the pulse is completed because every increased time step results in higher temperatures.

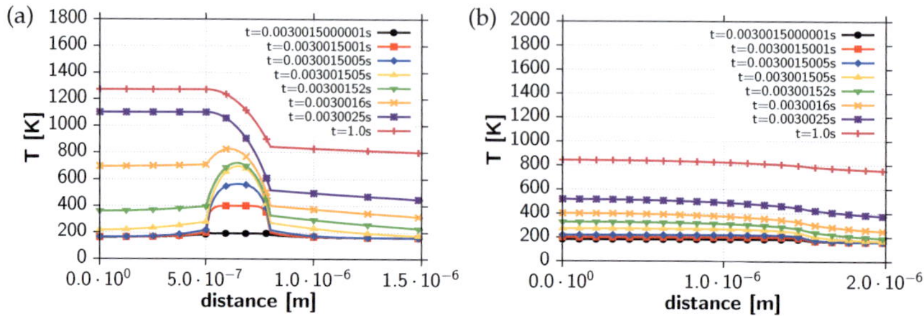

Figure 101: Temperature profiles for a free standing pillar with an initial temperature of $T_{init} = 160\,K$, a current density of $j = 1 \cdot 10^{10}\,A/m^2$ and a pulse duration of $dur_p = 1\,s$. The vertical cut of 98 is shown in (a) and the horizontal cut hc1 is shown in (b).

In contrast, an initial temperature of $T_{init} = 160\,K$ results in a different temperature profile [cf. Fig.101]. The temperature profiles exhibit a higher temperature increase in the center of the pillar within the first hundred nanoseconds [cf. Fig.101(a)].

To understand this behavior, the resistance dependent total power dissipation is shown in Fig.102 for the reference current density of $j = 1 \cdot 10^{10}\,A/m^2$. With respect to this behavior, the temporal temperature development can be explained. First, the initial electric resistance produces extremely high liberated power, even though the high temperature amplitude reduces the electric resistance almost simultaneously within the first nanoseconds. However, since PCMO has poor thermal conductivity, a tremendous amount of heat remains in the pillar within this short time period. Thus here, a maximum temperature gradient exists from the center of the PCMO material towards the top and bottom electrode until heat conduction towards these electrodes achieves significant influence. This temporal maximum gradient is especially pronounced at $t = 0,0030015005$ s in Fig.101(a).

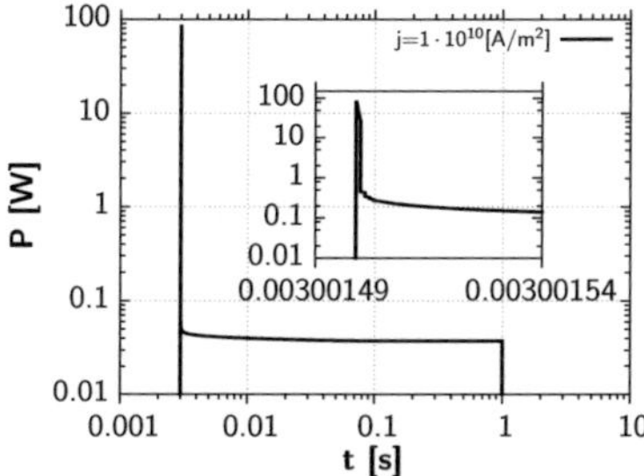

Figure 102: Resistance dependent total power dissipation within a 1 s pulse in a free standing pillar at T_{init} = 160 K.

From this time point, the temperature increases especially in the top electrode while the temperature in the bottom electrode remains always lower.   Hence, the temperature gradient from the center of the pillar towards the top decreases and finally turns around. Concurrently, tremendously less power is liberated since the electric resistance decreased significantly [cf. Fig.102]. However, there is still enough power released, to increase the temperature further. Especially towards the end of the 1 s pulse, a significant average temperature increase of the Pt bottom electrode and the substrate is observable. Comparing the time point t = 1.0 s to the time point t = 0.0030025 s in Fig.101 exhibits this fact. Based on this fact, a bigger temperature gradient exists between the top and bottom electrode after 1 µs in comparison to 1 s. Consequently, the mean PCMO temperature must be smaller after 1 µs than after 1 s.

Subsequently, the pad device with the electrically insulating PMMA layer is studied [cf. Fig.94(d)]. Here, just a pulse duration of dur_p = 1 s and an initial temperature of T = 160 K is considered. Apparently, a direct comparison of Fig.101(a) and Fig.103(a) exhibits a similar system behavior of a free standing pillar and the PCMO pad device.

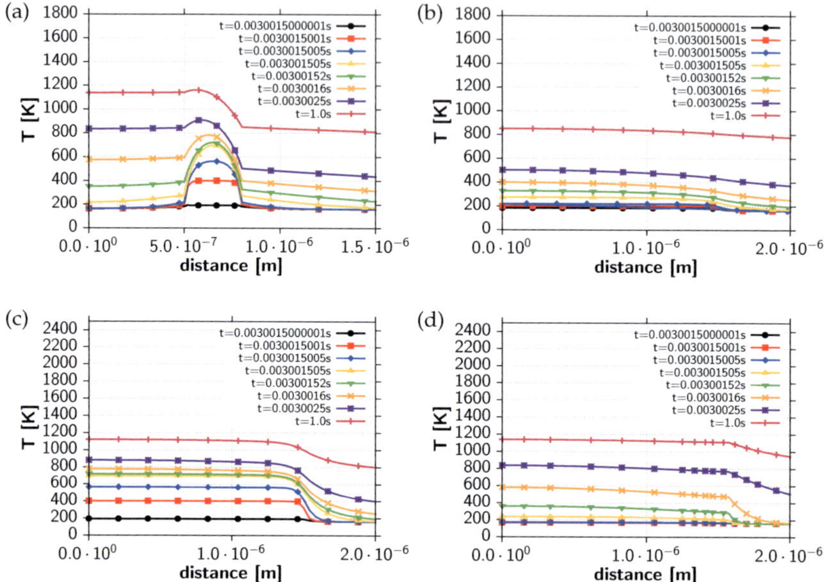

Figure 103: Temperature profiles for a PCMO layer with insulating PMMA material for an initial temperature of $T_{init} = 160$ K, a pulse duration of $dur_p = 1$ s and a current density of $j = 1 \cdot 10^{10}$ A/m^2. The vertical cut of 98 is shown in (a), the horizontal cuts hc1, hc2 and hc3 are shown in (b), (c) and (d) respectively.

However, the free standing pillar exhibits an approximately 100 K higher maximum temperature in comparison to the PCMO pad device. This circumstance is originated in the bigger spatial dimension of the covering Pt electrode and the adjacent PMMA material. Because here, induced heat spreads through the highly-conductive Pt electrode and additional thermal mass ($\rho \cdot c_p$) of the Pt electrode and the PMMA material has to be saturated. To illustrate these circumstances, the temperature profiles are given for the three different cuts hc1, hc2 and hc3 of Fig.98 in Fig.103(b), 103(c) and 103(d) respectively.

The horizontal cut hc1 shows, that for every time point, almost no temperature gradient exists from the axis of symmetry towards the side. On the contrary, the horizontal cut hc2 exhibits a significant temperature decrease from the actively heated PCMO material to the non actively heated PCMO material. Moreover, horizontal cut hc3 shows, that already after 1 µs of heating, a significant temperature gradient exists within the Pt top electrode. Thus, every horizontal cut illustrates that heat is spread to the side. **Consequently, the mean temperature in the PCMO layer is less than for the free standing pillar and hence, the electric resistance must be slightly higher for the PCMO pad device in comparison to the free standing pillar.** A corresponding temperature visualization of Fig.103 is given in Fig.104. This visualization helps to understand the previously mentioned temperature developments and completes the picture of heat spread.

(a)   (e)

(b)   (f)

(c)   (g)

(h)   (k)

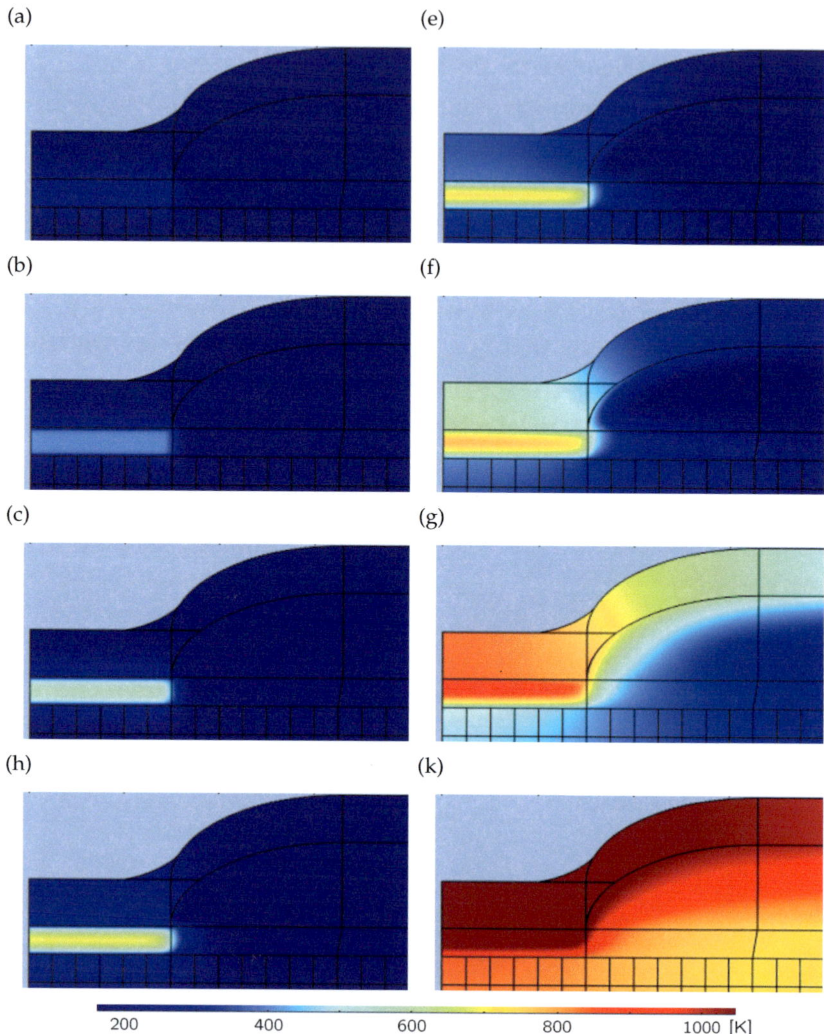

Figure 104: Temperature visualization for the PCMO pad device with insulating PMMA material [cf. Fig.94(b)]. The conditions are: Initial temperature of $T_{init} = 160$ K, a pulse duration of $dur_p = 1$ s and a current density of $j = 1 \cdot 10^{10}$ A/m^2. Shown are the respective time steps of Fig.103.

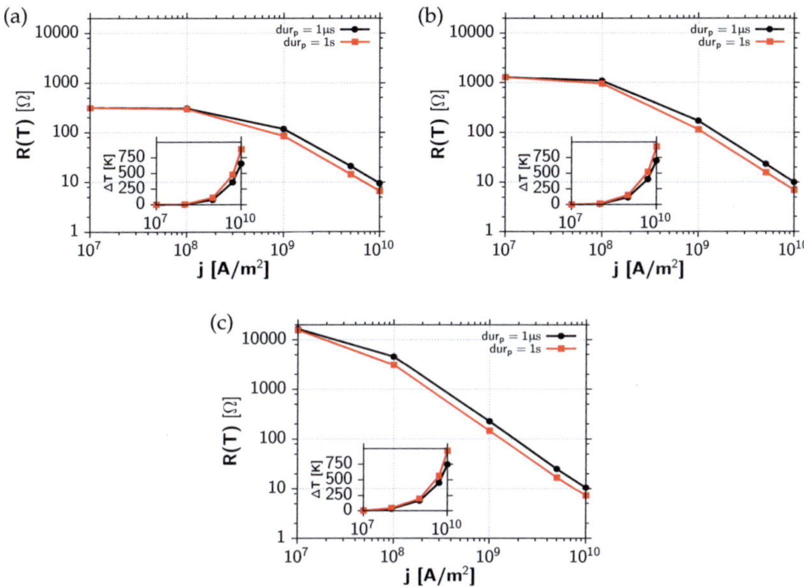

Figure 105: Temperature dependent resistance behavior for the free standing PCMO pillar without the Kleindiek contact [cf. Fig.94(c)]. The ambient temperatures are (a) 300 K, (b) 230 K and (c) 160 K. The insets contain the relative temperature increase ΔT within the heating pulse.

As result, the respective resistance behavior is shown in Fig.105 for the free standing PCMO pillar without the Kleindiek contact [cf. Fig.94(c)] as example. The electric resistance of the pillar is calculated by Eq.4.21 with the maximum occurring mean temperature, which we find at the end of the induced electric pulse [cf. Fig.97].

First, the initial temperature of T_{init} = 300 K is considered. The electric resistance remains almost the same until a current density of j = $1 \cdot 10^8$ A/m^2 is applied. From then on, the electric resistance decreases extremely with increasing current density. Essentially, the decrease in electric resistance occurs when the temperature also increases significantly.

The respective temperature amplitudes are indicated in the insets of Fig.105. This tendency appears for both, a 1 µs and 1 s induced current pulse. However, in general, the resistance always stays slightly higher for a 1 µs than for a 1 s pulse. **This implies in turn, that the major heating takes place within** 1 µs. Second, the initial temperature of T_{init} = 230 K is considered. Similar to an initial temperature of T_{init} = 300 K, the electric resistance remains almost the same for small current densities. However, the decrease in electric resistance already occurs for smaller current densities in comparison to the initial temperature of T_{init} = 300 K. Third, the initial temperature of T_{init} = 160 K is considered. Here, the electric resistance already decreases for small current densities in comparison to the previous initial temperatures. **As result, the electric resistance decreases tremendously for every initial temperature at high current densities.** A numerical comparison of the different temperature dependent electric resistivity values is given in Tab.4 for the different pillar configurations and the pad device [cf. Fig.94]. In general, the pillar configurations and the pad device exhibit the same temperature dependent resistivity behavior with increasing current densities. However, evidently the pad device always results for increasing current densities in the highest remaining electric resistance afterwards the respective pulses were applied.

As an overall result, the previous given simulations prove that extremely high temperatures occur for high current densities. Consequently, is seems plausible that such high temperatures could initiate oxide migration and hence assist the switching mechanism thermally.

T_{init} [K]	j [A/m^2]	Pillar cf. Fig.94(c)		Pillar with tungsten (W) contact cf. Fig.94(c)		Pad cf. Fig.94(d)	
		ΔT [K]	R(T) [Ω]	ΔT [K]	R(T) [Ω]	ΔT [K]	R(T) [Ω]
300	$1 \cdot 10^7$	0.04	312.06	0.04	312.06	0.03	312.09
	$1 \cdot 10^8$	3.92	293.73	3.92	293.78	3.35	296.30
	$1 \cdot 10^9$	111.67	84.39	111.76	84.37	102.55	91.51
	$1 \cdot 10^{10}$	884.45	6.62	884.61	6.62	817.81	7.26
230	$1 \cdot 10^7$	0.17	1290.66	0.17	1290.67	0.14	1291.53
	$1 \cdot 10^8$	12.58	947.54	12.60	947.92	11.08	982.04
	$1 \cdot 10^9$	150.45	113.13	150.46	113.08	140.47	125.41
	$1 \cdot 10^{10}$	924.04	6.90	924.07	6.90	857.12	7.61
160	$1 \cdot 10^7$	2.09	15676.86	2.09	15680.41	1.80	15911.98
	$1 \cdot 10^8$	40.89	3093.03	40.89	3091.73	38.09	3407.32
	$1 \cdot 10^9$	196.86	145.67	196.85	145.67	186.13	165.19
	$1 \cdot 10^{10}$	965.32	7.18	965.41	7.18	898.10	7.96

Table 4: Comparison of the different occurring temperature amplitudes and resulting resistances within a 1 s heating pulse for the different PCMO pillars [cf. Fig.94].

5 Methodology to determine the isotropic and anisotropic thermal conductivity

The analysis of the previously investigated geometry structures exhibits to some extent completely different ΔT-line behaviors in comparison to classic 3ω-method configurations. Thus, the offset model and slope methods are no longer applicable to determine the thermal conductivity since they are based on analytic descriptions which are no longer valid for these systems. Hence, a new method is required, which can be applied on various geometries, frequencies and material combinations. In this work, a new methodology is presented which combines the advantages of different structures and frequencies to determine the isotropic and anisotropic thermal conductivity.

5.1 General methodology

This methodology involves experiments and numerical analysis for the investigated materials. The general procedure is shown in Fig.106. After the sample is fabricated, a 3ω-measurement is performed for the investigated material first. Second, the presented Finite Element Model is used to simulate the exact given geometry configuration. The Finite Element simulations obtain the temperature amplitude ΔT as output. Third, the Neural Network is established. The Neural Network has to be trained in the range of the expected thermal conductivity. Moreover, the more dependencies are known, the better is the solution of the Neural Network in the end. In this work, the calculated ΔT for a given thermal conductivity is used to train a Neural Network. Essentially, the Network is trained in the opposite direction which means the Network learns the dependencies for given ΔT 's to κ.

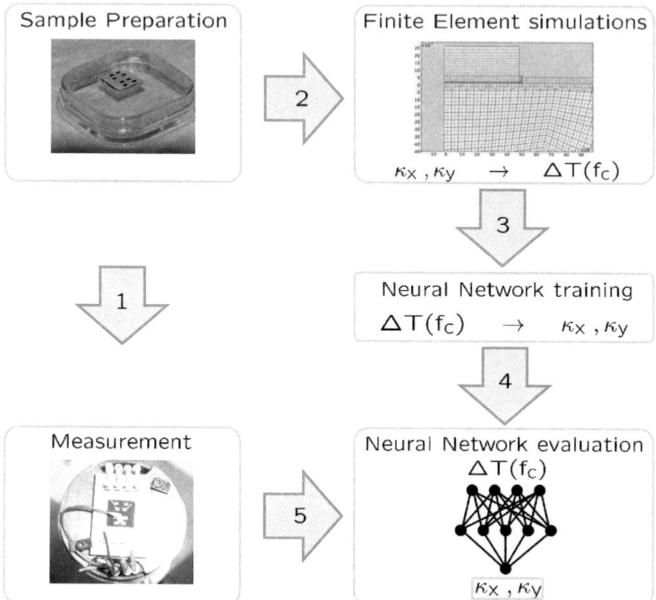

Figure 106: Illustration of the general procedure to determine thermal conductivities. The 'Sample Preparation' and 'Measurement' pictures are taken from the measurement setup, established by our project partners.

Since the Network can interpolate, it is able to solve the Inverse Problem. However, as shown in previous sections, anisotropic thermal conductivity variations can result for some geometry configurations in different characteristics in the ΔT-lines. An example for this behavior is given in Fig.86(b) for the long heater configuration of the 6-pad heater structure. Comparing in this plot the black and blue ΔT-lines exhibits, that the black line is above the blue line for low frequencies but turns around for high frequencies. If we assume that only one anisotropic thermal conductivity combination can exist for a clear defined ΔT-line, this configuration of the 6-pad heater structure can be used to determine the anisotropic thermal conductivity within one step (κ_x, κ_y). Therefore, this work applies a new approach by using the information of different ΔT-lines in general over a defined frequency regime instead of either the slope or the axis intercept. However, some configurations are not suitable to determine the cross- and in-plane thermal conductivity within one step. In this case, still two measurements and two Neural Networks are needed to identify the cross- and in-plane thermal conductivity separately because the respective sensitivity is too small. The last step combines the temperature amplitude, measured in experiments with the established Neural Networks to identify the correspondent thermal conductivities. By this procedure, we do not need to employ any restricting assumptions that may be needed to derive analytic solutions. Consequently, we can investigate many system configurations.

5.2 Neural Network and Inverse Problem

A Neural Network constitutes a mathematical instrument to represent and approximate arbitrary dependencies where no explicit analytical expressions are available [cf. 73, 140]. Its definition is originated in natural nervous systems and brains of humans and animals. Information of nervous systems are processed by a big number of simple and parallel working units [96]. These units are so-called Neurons. Whether information is exchanged or not, depends on activation signals.

Just as a human brain must learn dependencies in general, a Neural Network must be trained on a certain data basis before an outcome can be predicted.

The amount of data, the formatting and eventual incorporated mathematical background limits the 'knowledge and accuracy' of the Neural Network. Subsequent to the training process, the Neural Network is able to predict the previously unknown value as outcome. The Neural Networks which we apply in this work are developed by S. Scherrer-Rudiy [140] and especially N. Huber [73]. Just as this work focuses on thermal conductivities of thin layer materials, the fundamental work of N. Huber was used to identify mechanical properties of thin layer materials in [73] and mechanical properties in general by E. Tyulyukovskiy [154] and S. Scherrer-Rudiy [140].
In the following, the working principle of the Neural Network is described. The mathematical treatment of a Neuron is shown in Fig.107.

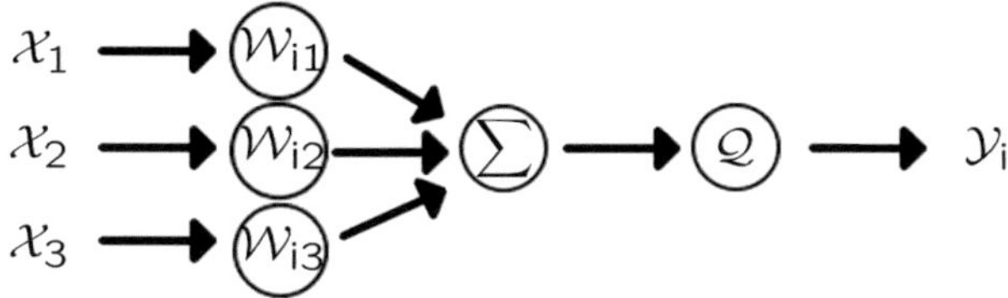

Figure 107: Illustration of the mathematical modeled Neuron.

Here, X_j is the input vector of the regarded Neuron. The entities of the input vector are multiplied by the synaptic weights W_{ij} and summed up. The activation of the Neuron, and thus whether or not the output vector y_i is greater than zero, depends finally on the activation function Q. Possible activation function examples are step functions or tangent hyperbolicus. A Neural Network contains several of such Neurons in different layers. In this work, a conventional Feed-Forward Network is used to describe the dependencies between temperature amplitude and thermal conductivities.[42] The Neural Network scheme is shown in Fig.108.

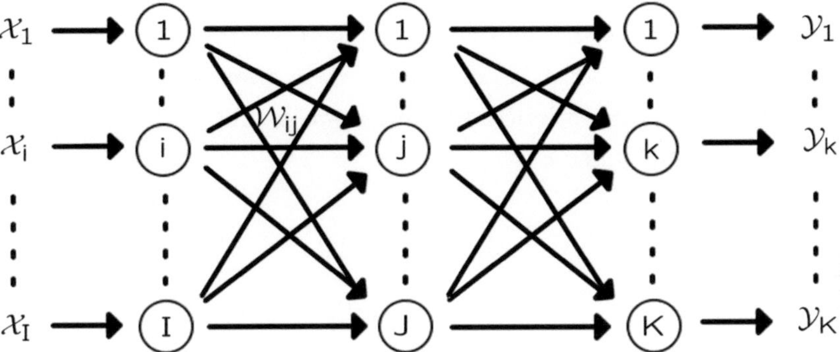

Figure 108: Illustration of a general Neural Network scheme with three different layers.

A Neural Network must contain at least one input and one output layer. Existing layers in between are so-called 'hidden layers'. While the number of in- and output values in the respective vector are predefined by the problem which has to be solved, the numbers of hidden layers and internal Neurons are subject to change. In general, more hidden layers and internal Neurons are necessary if the dependencies are complex.

Neural Networks learn the dependencies of given training examples by adapting the synaptic weights $\mathcal{W}$ so that the given training example is reproduced. The training examples contain a vector pair with the input and output values. The training process uses an optimization procedure for the object function

$$S(\mathcal{W}) = \mathcal{G}(\mathcal{W}) + \mathcal{E}(\mathcal{W}).\tag{5.1}$$

---

[42] A Feed-Forward Network contains only relations from the previous layer to the next layer. This means, no back loop exists from the next layer to the previous layer [96].

The object function $\mathcal{S}(\mathcal{W})$ consists of an error term $\mathcal{G}(\mathcal{W})$ and a regulation term $\mathcal{E}(\mathcal{W})$.[43]   The aim of the optimization is to find optimum synaptic weights, so that the difference between the output vector and the training example is as small as possible. By minimizing the object function, the Neural Network is trained.   Therefore, a gradient method is applied for an iterative improvement of the synaptic weights $\mathcal{W}$ so that

$$\mathcal{W}^{\mathcal{T}+1} = \mathcal{W}^{\mathcal{T}} - \eta \nabla \mathcal{S}(\mathcal{W}) . \qquad (5.2)$$

Here, $\mathcal{W}^{\mathcal{T}+1}$ is the next iterative synaptic weight, $\mathcal{W}^{\mathcal{T}}$ is the actual synaptic weight, η is an empirically determined scaling parameter and $\nabla \mathcal{S}(\mathcal{W})$ is the applied gradient on the object function $\mathcal{S}(\mathcal{W})$.

After sufficient iterative improvements, an optimum $\mathcal{W}$ is obtained for which the Neural Network has a minimum error.[44] However, the training requires caution, since too much training or a local minimum falsifies the result. Thus, the Neural Network has to be subsequently validated.   Therefore, a known data set for in- and output vectors must be excluded from training and checked with the established Neural Network afterwards.   Finally, the Neural Network can be used to predict the desired unknown values. In general, the quality of the result depends primarily on the quantity and quality of the used training examples. This means first: A sufficient fine pattern of data points (vectors) must be given for the training process. Second, the Network's prediction towards the borders of the numerical data contains a higher uncertainty since these data points are surrounded by less training data points.

---

[43] Detailed information about the object function is given by S. Scherrer-Rudiy [140]
[44] Detailed information about the error is given by S. Scherrer-Rudiy [140].

Third, predictions beyond the trained data basis are rather less reliable. At last, the quality of the Network decreases for wrong vector pairs, if they are used for training.[45]

In this work, Neural Networks are trained in a way, that they learn the thermal conductivities corresponding to the temperature amplitudes, so that

$$\mathcal{X}(\Delta T) \rightarrow \mathcal{Y}(\kappa). \tag{5.3}$$

Thus, the Network can use the temperature amplitudes which are measured in experiments and interpolate the corresponding thermal conductivities.

---

[45]Detailed descriptions and discussions about the quality of Neural Networks is given in [73], [140] and [96].

6 Application examples

6.1 STO layer on an STO substrate

In this section, the developed methodology is applied to determine the thermal conductivity of a thin STO layer in the top down geometry [cf. Fig.23]. Here we are interested in stress-strain dependent thermal conductivity variations. However, applying different materials as layer and substrate induces stresses into the thin layer by point defects and different lattice parameters [164]. Because of this, we reduce the degree of freedom of influence factors by applying the same substrate material such as the investigated layer material. Hence, the STO layer is placed on top of an STO substrate. The substrate is single crystalline. This allows to study homoepitaxial grown layers where the occurring compressive in-plane strain is purely induced by point defects [164]. The defects are incorporated by different thermal fabrication steps [cf. 164].

However, if two similar conducting materials are placed above each other, the offset model of Cahill [28] is incapable to determine the thermal conductivity. According to this problem, we apply our developed methodology to determine the isotropic thermal conductivity of several 300 nm thick pre-stressed STO layers. It has to be pointed out, that the layers are only pre-stressed in experiments and not in the Finite Elemente simulations.

In the following, a ΔT-line is defined by four frequency points $f_c = 40, 60, 80, 100$ Hz [cf. Fig.110(a)]. For these respective frequencies, measurements and simulations are processed [cf. Fig.106]. Afterwards, the data basis of the Finite Element simulations is used to train a Neural Network. Several Neural Networks have to be established, since different conditions exist for different measurements.

The varying conditions are: The power per length and heater's height vary between $P/L = 5 - 7.5$ W/m and $a = 100 - 300$ nm respectively. The thermal conductivity of the substrate varies between different STO samples and varies especially for different ambient temperatures. The heat capacity of the STO layer and the substrate also changes with different ambient temperature [51]. On the contrary, the heater's width $2b = 10$ μm and the substrate thickness $d_s = 1$ mm is the same throughout every experiment. Consequently, various Finite Element simulations are realized, to take the previously given conditions into account. In this application study, only the temperature field is considered. While we assume the boundaries Γ_{1-3} to be adiabatic, we take into account a heat transfer from the substrate to the mount at boundary Γ_4 by Eq.4.4 [cf. Fig.109]. Moreover, all materials are assumed to be isotropic. Hence, the partial differential equation for the substrate and for the heater are given by Eq.3.1 and Eq.3.2 respectively. The partial differential equation for the investigated layer material writes to

$$\Omega_{\text{lay}} : \qquad \frac{\partial^2 T_{\text{lay}}}{\partial x^2} + \frac{\partial^2 T_{\text{lay}}}{\partial y^2} - \frac{1}{D_{\text{lay}}} \frac{\partial T_{\text{lay}}}{\partial t} = 0 . \qquad (6.1)$$

The training of the Neural Network is done first by declaring four temperature amplitude values as an input vector. These values are then correlated to the output vector, which is the thermal conductivity. A schematic overview is given in Fig.110(b). Here, the training applies two hidden layers and between 5 to 30 internal Neurons. It turned out that the object function $S(\mathcal{W})$ can be minimized more efficiently by multiplying the ΔT values by 10, so that the input vectors $X(\Delta T)$ and output vectors $\mathcal{Y}(\kappa)$ are on the same order of magnitude.

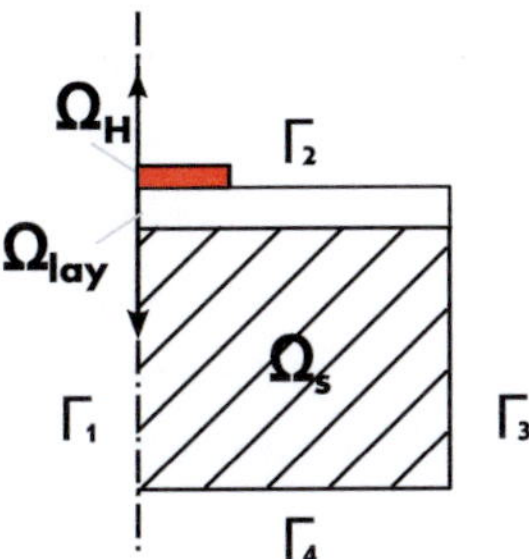

Figure 109: Considered system for the Finite Element simulation.

In this study, we investigate the thermal conductivity at room temperature T_0 = 300 K for different pre-strained layers first. An example for several simulated data pairs is plotted together with the respective measurement in Fig.110(c) for a strain of ε = 0,62%. The result of the Neural Network identification for the corresponding thermal conductivity is shown in Fig.111(a). Moreover, this figure contains the results for different pre-strained layers down to ε = 0,17%. This study at room temperature exhibits, that the thermal conductivity decreases from approximately 10 W/mK at ε = 0,17 % to approximately 5 W/mK at ε = 0,62 %. Hence, the thermal conductivity is reduced by 50% and thus decreases significantly.

Second, the temperature dependence on pre-strained layers is investigated. An example for several simulated data pairs is plotted together with the respective measurement in Fig.110(d) for a temperature of T_0 = 200 K and a strain of ε = 0,62 %. The result for the corresponding thermal conductivity is shown in Fig.111(b) together with the results for measuring temperatures down to 40 K.

In this figure, it is shown that, while the thermal conductivity of the unstrained substrate increases slightly towards a temperature of 40 K, the thermal conductivity of the thin STO layer decreases further. Here, a thermal conductivity around $\kappa \approx 1$ W/mK is determined at 40 K. **As an overall consequence, the previous results indicate a tremendous impact of strain on the thermal conductivity of thin STO layers.** For detailed explanation, see Wiedigen et al.[164].

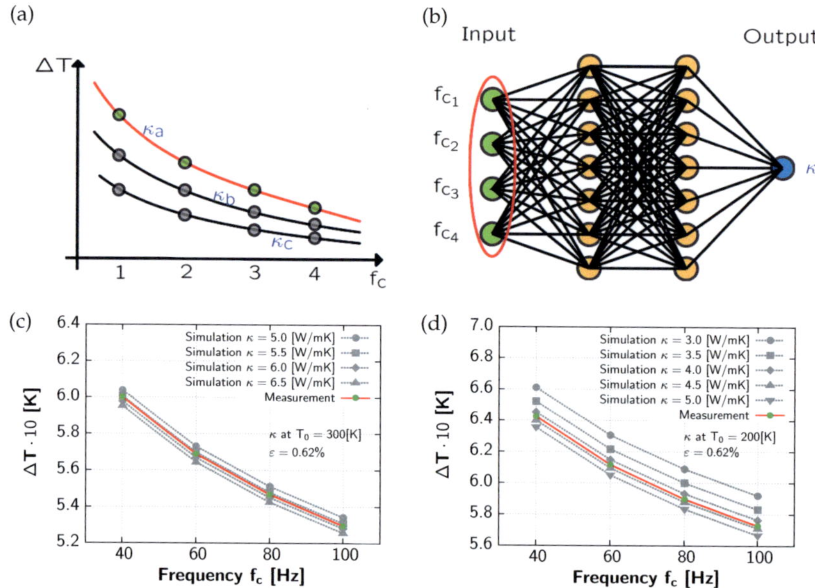

Figure 110: (a) Schematic assignment of thermal conductivities, based on a temperature amplitude over frequency plot. (b) Sketch of the Neural Network for the thermal conductivity determination. (c) Example of the line identification for a measurement at a strain of $\varepsilon = 0,62$ % within the numerical simulation data base values for an ambient temperature of $T_0 = 300$ K and an identified thermal conductivity of $\kappa_{lay} = 5.48$ W/mK and (d) an identified thermal conductivity of $\kappa_{lay} = 4.27$ W/mK at $T_0 = 200$ K.

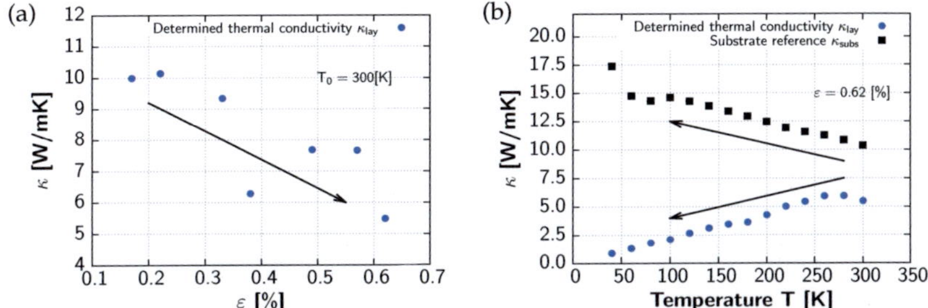

Figure 111: (a) Results for the determined strain dependent thermal conductivity at $T_0 = 300$ K. (b) Results for the determined strain and temperature dependent thermal conductivity at $\varepsilon = 0,62$ % in comparison to the applied and unstrained STO substrate.

6.2 Bottom electrode geometry

In this section, the bottom electrode geometry is used in combination with the developed methodology to determine the thermal conductivity of a thin poorly-conductive STO layer. However, as shown in Sec.4.2.2, a heater substrate platform is rather insensitive for poor conductive layers on top. Due to this problem, we apply here a heat sink on top of the layer material to increase the sensitivity for the 3ω-method to determine the thermal conductivity of the poorly-conductive layer. A scanning electron microscopy image is shown in Fig.112 for the bottom electrode geometry configuration which we investigate in this section.

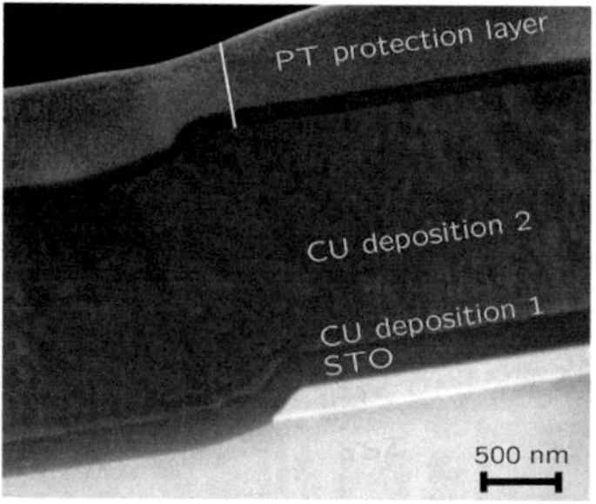

Figure 112: SEM image of the bottom electrode geometry. The SEM image was taken by our project partners of the University of Göttingen. The Pt layer on top of the Cu sink is not present for the 3ω-measurement. It exists only to protect the bottom electrode geometry for the SEM imaging process.

In the following, the manufacturing and measurement process are described first and afterwards, the numerical simulations are explained.

First, the heater substrate platform is fabricated and a 3ω-measurement is performed to determine the thermal conductivity of the substrate [cf. Fig.113(a)]. Here, a Pt heater with a height of $a = 203$ nm and a width

of 2b = 10 μm is applied on top of a 1 mm thick YSZ substrate.[46]
Second, the thin STO layer is placed on top of the heater substrate platform and a 3ω-measurement is performed [cf. Fig.113(b)]. Here, the thin STO layer on top is at least as wide, as the substrate is thick. Hence, the STO layer is considered as full covering. The thickness of the STO layer is $d_{lay} = 221$ nm.

Third, the Cu heat sink is placed on top of the layer by two deposition steps. Afterwards, a 3ω-measurement is performed for this stacked material combination [cf. Fig.113(c)]. Due to the fabrication process, the Cu heat sink on top has a width of $2b_{sin} = 500$ μm and a height of $d_{sin} = 1.454$ μm. In this section, every measurement and simulation applies a power per length of P/L = 5 W/m and measuring frequencies of $f_c = 100, 200, 400, 600, 800, 1000$ Hz. Fig.114 contains the temperature amplitude measured in experiments at the defined frequency points for the pure heater substrate platform (Pt-YSZ), the heater substrate platform with the STO layer on top (STO-Pt-YSZ) and the heater substrate platform with the additional heat sink on top (Cu-STO-Pt-YSZ).

For this application example, the numerical analysis can be divided into two major steps. In the first step, the thermal conductivity of the substrate is determined, and in the second step, the thermal conductivity of the investigated layer material is identified.

To determine the thermal conductivity of the substrate, the pure heater substrate platform is considered [cf. Fig.113(a)]. For this system, Finite Element simulations generate a data basis for thermal conductivity variations of the substrate first. In this application study, only the temperature field is considered.

---

[46]In this section, every material placed on top of the heater substrate platform is fabricated by a sputtering process.

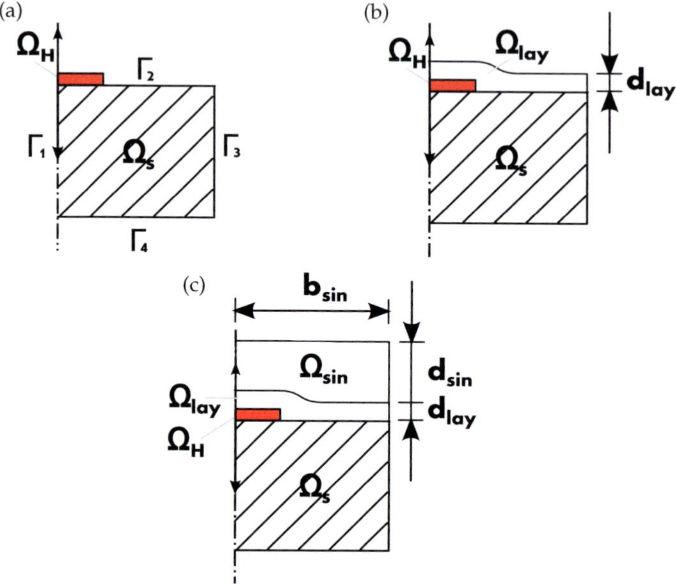

Figure 113: (a) Pure heater substrate platform; (b) thin layer on top of the heater substrate structure; (c) heat sink applied on top of the layer.

While we assume the boundaries Γ_{1-3} to be adiabatic, we take into account a heat transfer from the substrate to the mount at boundary Γ_4 by Eq.4.4 [cf. Fig.113(a)]. Moreover, all materials are assumed to be isotropic. Hence, the partial differential equation for the substrate and for the heater are given by Eq.3.1 and Eq.3.2 respectively.

Afterwards the Finite Element simulations are performed, the established data basis is used to train a Neural Network. The training of the Neural Network applies two hidden layers and 100 internal Neurons. Here, the training is done by declaring six temperature amplitude values as an input vector. These values are then correlated to the output vector, which is the thermal conductivity of the substrate [cf. Fig.115(b)].

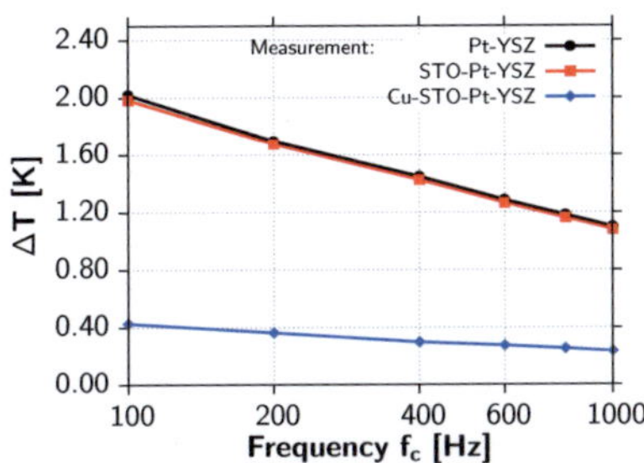

Figure 114: Temperature amplitudes measured in experiments at the defined frequency points for the pure heater substrate platform (Pt-YSZ), the heater substrate platform with the STO layer on top (STO-Pt-YSZ) and the heater substrate platform with the additional heat sink on top (Cu-STO-Pt-YSZ). Here, every measurement line is the arithmetic mean value of three different measurements at exactly the same sample.

Subsequently, the established Neural Network is used to solve the Inverse Problem and identify the thermal conductivity of the substrate from the temperature amplitudes measured in experiments. This identification is shown in Fig.115(c). **The identified thermal conductivity of the substrate is** $\kappa_s = 2.10$ **W/mK.**[47]
In the second step, the thermal conductivity of the layer material is identified. Therefore, Finite Element simulations are processed for the stacked material combination of Fig.113(c). Thus for the heater substrate platform with layer material and heat sink on top. Here, the previously determined thermal conductivity of the substrate is used. In this second step, the thermal conductivity of the investigated STO layer material is varied in the Finite Element simulations. Here, we also assume every outer boundary to be adiabatic except for the substrate mount, where we take into account again a heat transfer from the substrate to the mount.

---

[47]The literature value for the thermal conductivity of YSZ is $\kappa = 1.8$ W/mK.

Also here, every material is assumed to be isotropic. The partial differential equation for the substrate and for the heater are given by Eq.3.1 and Eq.3.2. The partial differential equations for the heat sink and the investigated layer material write to

$$\Omega_{sin}: \quad \frac{\partial^2 T_{sin}}{\partial x^2} + \frac{\partial^2 T_{sin}}{\partial y^2} - \frac{1}{D_{sin}}\frac{\partial T_{sin}}{\partial t} = 0, \tag{6.2}$$

$$\Omega_{lay}: \quad \frac{\partial^2 T_{lay}}{\partial x^2} + \frac{\partial^2 T_{lay}}{\partial y^2} - \frac{1}{D_{lay}}\frac{\partial T_{lay}}{\partial t} = 0. \tag{6.3}$$

Based on the Finite Element simulations, a second Neural Network is established. Again, the training of the Neural Network is done by declaring six temperature amplitude values as an input vector. These values are then correlated to the output vector, which is now the thermal conductivity of the layer material. Finally, this second Neural Network is used to solve the Inverse Problem and identify the thermal conductivity of the investigated STO layer material from the temperature amplitudes measured in experiments. The corresponding identification is shown in Fig.115(d). **The identified thermal conductivity of the layer is** $\kappa_{lay} = 1.2$ **W/mK.** This identified thermal conductivity of the STO layer seems rather small in comparison to the other STO layers of this work. However, in case of the bottom electrode, the placed STO layer is grown polycrystalline. Hence, we believe that the thermal conductivity should be significantly smaller. However, as shown in Sec.4.2.3, this bottom electrode geometry configuration is especially sensitive and suitable to determine poor cross-plane thermal conductivities. **Consequently, this identified thermal conductivity is the cross-plane thermal conductivity of the layer material. Moreover, this application example proofs that the the bottom electrode geometry can be used to determine the thermal conductivity of poorly-conductive layer materials in combination with the established methodology of this work.**

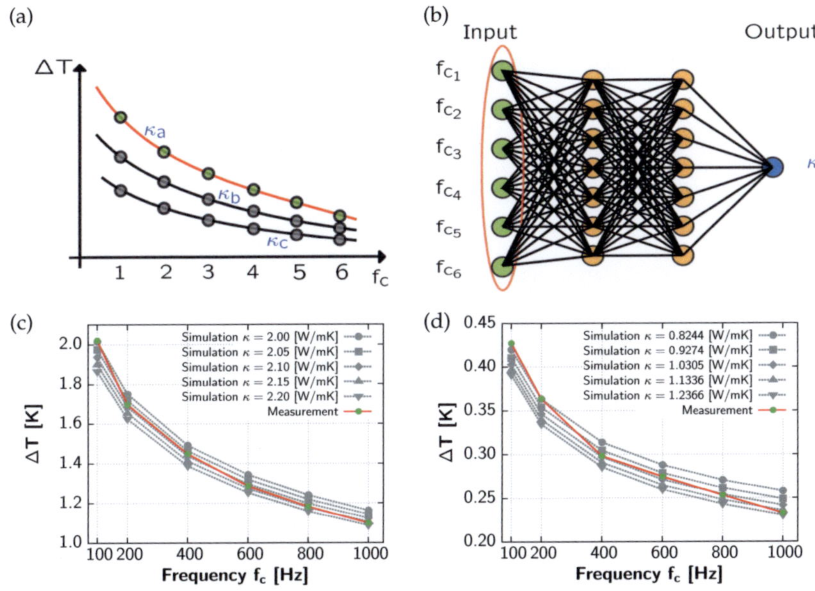

Figure 115: (a) Schematic assignment of thermal conductivities, based on a temperature amplitude over frequency plot. (b) Sketch of the Neural Network for the thermal conductivity determination. (c) Example of the line identification for the heater substrate platform measurement within the numerical simulation data base values for an identified thermal conductivity of $\kappa_s = 2.10$ W/mK and (d) an identified thermal conductivity of $\kappa_{lay} = 1.2$ W/mK for the investigated layer material.

7 Summary

This work aims to deepen the understanding of thermal conduction in nanostructured materials to design new thermoelectrics with higher efficiency than these materials exhibit today. However, therefore a measurement technique is required where we focus on the 3ω-method. Thus, the primary objectives of this work are to examine macroscopic influence factors within a 3ω-measurement, to develop new geometric measuring configurations in order to determine the isotropic and anisotropic thermal conductivity, and to establish a new evaluation methodology. Since thermoelectricity and the 3ω-method are based on heat propagation, the fundamentals of heat conduction are derived in the beginning.

The second section elucidates the concept, basic principles, and geometric configurations of the 3ω-method. Existing literature is reviewed and several solutions of the top down geometry are compared. The range of applicability is also examined. Moreover, fundamental work for the bottom electrode geometry is introduced and the potential use of a new heater substrate platform is presented.

The third section presents the governing equations of the used Finite Element Model. Here, the equations are derived in a partial differential sense. The interaction between the distinct macroscopic influence factors are explained and the respective consideration is motivated. This includes the temperature dependent electric resistivity of the heater, the electromagnetic induced skin effect, and the mechanical displacements within a measurement. Moreover, details for the time-dependent analysis are given and a mesh-block building system is presented to obtain a pre-defined mesh in order to achieve high precision analysis.

The fourth section applies the presented Finite Element Model on various geometric structures. Moreover, macroscopic influences, which are not commonly adressed in literature are examined. For instance, a temperature dependent electrical resistivity changes the temperature amplitude on the 4th decimal place. Taking into account thermal expansion can have an impact on the 3rd decimal place. On the contrary, the occurring skin effect has almost no influence on the temperature amplitude for frequencies up to 1 MHz.
Furthermore, mono- and multilayer configurations are investigated. First, analytic descriptions are compared to the Finite Element solution and differences are identified. In comparison to the given descriptions, the Finite Element solution appears to be the most precise solution at high frequencies. Hence, the performed simulations allow understanding the system behavior especially in the high frequency regime. Second, in order to determine the anisotropic thermal conductivity, different heater geometries are studied on top of mono- and multilayer configurations. Here, also non-classic layer substrate configurations are studied where poorly-conductive substrates are applied.
Additionally, a bottom electrode geometry is investigated to determine the isotropic and anisotropic thermal conductivity of nanoscale samples. Here the bottom electrode is especially examined for the anisotropic thermal conductivity determination of thin layers. This includes the application of heat sinks on top of the layer. In principal, the investigated heat sink configurations allow the determination of the anisotropic thermal conductivity. However,

since fabrication seems rather complicated, these configurations are inconvenient to determine the anisotropic thermal conductivity.

We think, that membrane structures are the solution to determine the anisotropic thermal conductivity. Therefore, these structures are investigated in this work. First, two-dimensional models are studied to understand the general system behavior. This includes different geometrical properties of the membrane, the applied power per length, and heat loss by radiation. Second, a three-dimensional model is developed in this work. The so-called 6-pad heater structure allows to determine for both materials, STO and PCMO, the in-plane thermal conductivity. Moreover, if a moderately-conductive membrane could be applied, the 6-pad heater structure would also provide the possibility to determine the in-plane thermal conductivity as well as the cross-plane thermal conductivity for an extremely poorly-conductive layer material.

Since micro- to nanoscale PCMO samples exhibit a change in electric resistance by electric stimulation (resistive switching), such as short impulse induced high current densities, this material has also been investigated in pillar geometries. Here the simulations provide a substantial contribution to understand the temperature development and thus the electric resistivity behavior in these experiments.

The fifth section presents a new methodology to determine the thermal conductivity, where no analytic solution is available. Since Finite Element simulations and experimental measurements result in temperature amplitudes, a procedure is necessary to merge simulations and experiments for the thermal conductivity determination. In this work, Finite Element simulations and Neural Networks are combined to evaluate experimental measurements.

Here, the Neural Networks are based on Finite Element results and connect experimental measurements with their respective simulated thermal conductivities. Hence, this allows to solve the Inverse Problem and thus determine the desired sample properties. The presented methodology can be applied to various geometric configurations such as the bottom electrode or the 6-pad heater structure. However, classic configurations can also be studied, where no offset model or slope method is applicable.

The last section is dedicated to two application example of the new methodology. The first example applies the developed methodology on a top down geometry. Here, a thin STO layer is investigated for stress-strain dependent thermal conductivity. To exclude an influence through the substrate by different lattice parameters of layer and substrate, the STO layer is deposed by a sputtering process on top of an STO substrate. Hence, a top down configuration exists, where no clear defined offset occurs. Based on the precise Finite Element simulations and the new developed methodology, it is possible to study the influence of strain on the thermal conductivity. As result, the thermal characterization exhibits a strong influence of strain on the thermal conductivity of thin STO layer material.

The second example applies the developed methodology on a bottom electrode geometry where a heat sink is applied on top of the investigated layer material. This example proofs the feasibility for a bottom electrode geometry to determine the thermal conductivity of a poorly-conductive thermoelectic material.

Overall, this thesis presents a comprehensive understanding on macroscopic factors, influencing Joule heating measurements at small scales. The simulations quantify the absolute temperature and temperature amplitude for structures used in 3ω-measurements and in pillar and pad structure experiments. Based on these simulations, new geometric configurations are developed for the 3ω-method and combined with a new methodology to determine the isotropic and anisotropic thermal conductivity. Finally, the developed evaluation methodology can be generalized and consequently be applied to a wider range of Joule heating applications for characterizing thermal properties of small scale materials.

Appendix A. Detailed information on formula

A.1 Modulated voltage

$$U = I \cdot R \tag{A.11}$$

$$= I_0 \cos(\omega t) \cdot [R_{av} + \Delta R \cos(2\omega t + \varsigma)]$$

$$= I_0 R_{av} \cos(\omega t) + I_0 \Delta R \cos(\omega t) \cos(2\omega t + \varsigma)$$

$$= I_0 R_{av} \cos(\omega t) + I_0 \Delta R \frac{1}{2} [\cos(\omega t - (2\omega t + \varsigma)) + \cos(\omega t + (2\omega t + \varsigma))]$$

$$= I_0 R_{av} \cos(\omega t) + I_0 \Delta R \frac{1}{2} [\cos(\omega t) \cos(2\omega t + \varsigma)$$

$$+ \sin(\omega t) \sin(2\omega t + \varsigma) + \cos(\omega t) \cos(2\omega t + \varsigma)$$

$$- \sin(\omega t) \sin(2\omega t + \varsigma)]$$

$$U = I_0 R_{av} \cos(\omega t) + \frac{I_0 \Delta R}{2} (\cos(3\omega t + \varsigma) + \cos(\omega t + \varsigma)) \tag{A.12}$$

A.2 Information about the approximate solution

Cahill [28] proposed an approximate solution for the case that the thermal penetration depth is large in comparison to the half heater width ($q^{-1} \gg b \rightarrow bq \ll 1$). He sets the limits of the integral from 0 to $1/b$ and assumes that

$$\frac{\sin^2(kb)}{(kb)^2} = 1.$$ (A.23)

Thus, Eq.2.17 writes now as:

$$\Delta T_h^C = \frac{P_0}{L\pi\kappa_s} \int_0^{1/b} \frac{1}{\sqrt{k^2 + q^2}} \, dk = \frac{P_0}{L\pi\kappa_s} \cdot \ln\left|k + \sqrt{k^2 + q^2}\right|_0^{1/b}$$

$$= \frac{P_0}{L\pi\kappa_s} \left(\ln\left|\frac{1}{b} + \sqrt{\frac{1}{b^2} + q^2}\right| - \ln|q|\right) = \frac{P_0}{L\pi\kappa_s}\left[\ln\left(\frac{\frac{1}{b} + \sqrt{\frac{1}{b^2} + q^2}}{q}\right)\right]$$

$$= \frac{P_0}{L\pi\kappa_s}\left[\ln\left(\frac{1}{bq} + \sqrt{\frac{1}{b^2q^2} + 1}\right)\right] = \frac{P_0}{L\pi\kappa_s}\left[\ln\left(\frac{1 + \sqrt{1 + bq^2}}{bq}\right)\right]$$

$$= \frac{P_0}{L\pi\kappa_s}\left[-\ln(bq) + \ln(2)\right] = \frac{P_0}{L\pi\kappa_s}\left[-\ln\left(b\left(\frac{D}{i2\omega}\right)^{-\frac{1}{2}}\right) + \ln(2)\right]$$

$$= \frac{P_0}{L\pi\kappa_s}\left[-\frac{1}{2}\ln\left(b^2\left(\frac{D}{i2\omega}\right)^{-1}\right) + \ln(2)\right] = \frac{P_0}{L\pi\kappa_s}\left[-\frac{1}{2}\ln\left(b^2\frac{i2\omega}{D}\right) + \ln(2)\right]$$

$$= \frac{P_0}{L\pi\kappa_s}\left[-\frac{1}{2}\ln(2\omega) - \frac{1}{2}\ln\left(\frac{ib^2}{D}\right) + \ln(2)\right]$$

$$= \frac{P_0}{L\pi\kappa_s}\left[-\frac{1}{2}\ln(\omega) - \frac{1}{2}\ln\left(\frac{ib^2}{D}\right) - \frac{1}{2}\ln(2) + \ln(2)\right]$$

$$\Delta T_h^C = \frac{P_0}{L\pi\kappa_s}\left[-\frac{1}{2}\ln(\omega) - \frac{1}{2}\ln\left(\frac{ib^2}{D}\right) + const\right]$$ (A.24)

Here, the constant reads to

$$\text{const} = -\frac{1}{2}\ln(2) + \ln(2) = \frac{1}{2}\ln(2). \tag{A.25}$$

Moreover, the temperature oscillation of the line heat source $\Delta T(r)$ from Eq.2.8 is also approximated in the following. If the distance r from the heat source is small in comparison to the thermal penetration depth, the modified Bessel function K_0^B can be approximated by neglecting terms of higher order so that K_0^B reads as [1, p. 375]

$$K_0^B(s) = -\left(\ln\frac{1}{2}s + \gamma\right) \cdot I_0^B(s) + \frac{\frac{1}{4}s^2}{(1!)^2} + \left(1 + \frac{1}{2}\right) \cdot \frac{(\frac{1}{4}s^2)^2}{(2!)^2} + \dots \tag{A.26}$$

$$= -\ln(s) + \ln(2) - \gamma. \tag{A.27}$$

Here, $\gamma = 0.577215\dots$ [37] is the Euler-Maschoneri constant and K_0^B involves the modified Bessel function of first kind and zero order I_0^B.[48]

---

[48] $I_0^B(s) = 1 + \frac{\frac{1}{4}s^2}{(1!)^2} + \dots$

With the latter approximation, the modified Bessel function K_0^B reads as

$$K_0^B(qr) = -\ln\left[\left(\frac{D}{i2\omega}\right)^{-\frac{1}{2}} \cdot r\right] + \ln(2) - \gamma \tag{A.28}$$

$$= -\frac{1}{2} \cdot \ln\left[\left(\frac{D}{i2\omega}\right)^{-1} \cdot r^2\right] + \ln(2) - \gamma$$

$$= -\frac{1}{2} \cdot \ln\left(\frac{r^2}{\frac{D}{i2\omega}}\right) + \ln(2) - \gamma$$

$$= -\frac{1}{2} \cdot \left[\ln\left(\frac{r^2}{D}\right) + \ln(i2\omega)\right] + \ln(2) - \gamma$$

$$= -\frac{1}{2} \cdot \left[\ln(r^2) - \ln(D)\right] - \frac{1}{2} \cdot \ln(i2\omega) + \ln(2) - \gamma$$

$$= \frac{1}{2} \cdot \left[\ln(D) - \ln(r^2)\right] - \frac{1}{2} \cdot \ln(i2\omega) + \ln(2) - \gamma$$

$$= \frac{1}{2} \cdot \ln\left(\frac{D}{r^2}\right) - \frac{1}{2} \cdot \ln(2\omega) - \frac{1}{2} \cdot \ln(i) + \ln(2) - \gamma$$

$$= \frac{1}{2} \cdot \ln\left(\frac{D}{r^2}\right) - \frac{1}{2} \cdot \ln(2\omega) - \frac{1}{2} \cdot \frac{\pi \cdot i}{2} + \ln(2) - \gamma$$

$$K_0^B(qr) = \frac{1}{2} \cdot \ln\left(\frac{D}{r^2}\right) - \frac{1}{2} \cdot \ln(2\omega) + \ln(2) - \gamma - \frac{\pi \cdot i}{4} \tag{A.29}$$

With the latter approximation of the modified Bessel function, the temperature oscillation of Eq.2.8 writes to

$$\Delta T(r) = \frac{P_0}{L\pi\kappa_s}\left[\frac{1}{2} \cdot \ln\left(\frac{D}{r^2}\right) - \frac{1}{2} \cdot \ln(2\omega) + \ln(2) - \gamma - \frac{\pi \cdot i}{4}\right]. \tag{A.210}$$

A.3 Definition of the skin depth and derivation of decoupled equations for the magnetic and electric field intensity

First, the skin depth is introduced. The quantity of the skin depth describes the decrease of the current density with increasing distance from a surface of a material into it [130]. Per definition [130, p. 385], the skin depth δ_{skin} m writes to

$$\delta_{\text{skin}} = \sqrt{\frac{2}{\omega \sigma_{\text{ec}} \mu}}.$$ (A.311)

For high frequencies, the current is restricted to an extremely thin layer near the conductors surface [130].

Second, the decoupled equations for the magnetic and electric field intensity are derived. Therefore, the following vector identity [cf. 76] is given under consideration of Eq.3.21:

$$\nabla \times \nabla \times \mathbf{H} = \nabla(\nabla \cdot \mathbf{H}) - \nabla^2 \mathbf{H}$$ (A.312)

$$= \nabla(\mu^{-1} \cdot \nabla \cdot \mathbf{B}) - \nabla^2 \mathbf{H}$$

$$= -\nabla^2 \mathbf{H}$$ (A.313)

To get the wave equation, one can take the curl of the extended Ampère's law [Eq.3.18]. Thus, we get with the constitutive relations:

$$\nabla \times \nabla \times \mathbf{H} = \nabla \times \frac{\partial \mathbf{D}}{\partial t} + \nabla \times \mathbf{j} \qquad (A.314)$$

$$-\nabla^2 \mathbf{H} = \nabla \times \left(\epsilon \frac{\partial \mathbf{E}}{\partial t}\right) + \nabla \times (\sigma_{ec} \cdot \mathbf{E}) \qquad \Big| \quad \text{cf. Eq.3.31, 3.29}$$

$$= \epsilon \frac{\partial (\nabla \times \mathbf{E})}{\partial t} + \sigma_{ec}(\nabla \times \mathbf{E})$$

$$= -\epsilon \left(\frac{\partial^2 \mathbf{B}}{\partial t^2}\right) + \sigma_{ec}\left(-\frac{\partial \mathbf{B}}{\partial t}\right) \qquad \Big| \quad \text{cf. Eq.3.20}$$

$$= -\epsilon\mu \left(\frac{\partial^2 \mathbf{H}}{\partial t^2}\right) + \sigma_{ec}\mu\left(-\frac{\partial \mathbf{H}}{\partial t}\right) \qquad \Big| \quad \text{cf. Eq.3.30}$$

$$\nabla^2 \mathbf{H} = \epsilon\mu \left(\frac{\partial^2 \mathbf{H}}{\partial t^2}\right) + \sigma_{ec}\mu\left(\frac{\partial \mathbf{H}}{\partial t}\right) \qquad (A.315)$$

Equivalent to the previous equation, a similar derivation can be done for Eq.3.20, the Faraday's law, to derivate

$$\nabla^2 \mathbf{E} = \epsilon\mu \left(\frac{\partial^2 \mathbf{E}}{\partial t^2}\right) + \sigma_{ec}\mu\left(\frac{\partial \mathbf{E}}{\partial t}\right), \qquad (A.316)$$

for the electric field intensity. Thus both, the magnetic and electric field intensity can be decoupled but still contain the information about the electro-magnetic waves. Considering the first and second time derivative [cf. Sec.3.1.2], $i\omega$ and $-\omega^2$ respectively, Eq.A.315 and Eq.A.316 develop to:

$$\nabla^2 \mathbb{H} = (-\omega^2\epsilon\mu + i\omega\sigma_{ec}\mu)\mathbb{H} \qquad (A.317)$$

$$\nabla^2 \mathbb{E} = (-\omega^2\epsilon\mu + i\omega\sigma_{ec}\mu)\mathbb{E} \qquad (A.318)$$

This pair of equations are the so-called complex Helmholtz-equations. They can be written as

$$-\nabla^2 \mathbb{H} + k^2 \mathbb{H} = 0, \qquad (A.319)$$

$$-\nabla^2 \mathbb{E} + k^2 \mathbb{E} = 0, \qquad (A.320)$$

where k^2 describes the complex wave. If terms of higher order are neglected, as mentioned in Sec.3.1.2, k^2 reduces to

$$k^2 = i\omega\sigma_{ec}\mu, \qquad (A.321)$$

and thus

$$k = \sqrt{i\omega\sigma_{ec}\mu}. \qquad (A.322)$$

With the mathematical identity $\sqrt{i} = (1 + i)/\sqrt{2}$ we get

$$k = \sqrt{\omega\sigma_{ec}\mu} \cdot (1 + i) \cdot \frac{1}{\sqrt{2}} = \sqrt{\frac{1}{2} \cdot \omega\sigma_{ec}\mu} \cdot (1 + i). \qquad (A.323)$$

With the skin depth definition of Eq.A.311, Eq.A.323 becomes

$$\sqrt{\frac{1}{2} \cdot \omega\sigma_{ec}\mu} \cdot (1 + i) = \frac{1}{\delta_{skin}}(1 + i). \qquad (A.324)$$

Consequently, the information of the skin depth is present in both decoupled complex Helmholtz-equations.

A.4 Derivation for the complex Helmholtz-Equation

$$\nabla \times \mathbb{H} = i\omega \mathbb{D} + \mathbb{J} \tag{A.425}$$

$$\nabla \times \mathbb{H} = i\omega \mathbb{D} + \sigma \cdot (-i\omega \mathbb{A}_j)$$

$$\nabla \times \mathbb{H} = i\omega \epsilon \cdot \mathbb{E} + \sigma \cdot (-i\omega \mathbb{A}_j)$$

$$\nabla \times \mathbb{H} = i\omega \epsilon \cdot (-i\omega \mathbb{A}_j) + \sigma \cdot (-i\omega \mathbb{A}_j)$$

$$\nabla \times \mathbb{H} = \omega^2 \epsilon \cdot \mathbb{A}_j + \sigma \cdot (-i\omega \mathbb{A}_j)$$

$$\nabla \times \frac{\mathbb{B}}{\mu} = \omega^2 \epsilon \cdot \mathbb{A}_j + \sigma \cdot (-i\omega \mathbb{A}_j)$$

$$\nabla \times \mu^{-1} \nabla \times \mathbb{A}_j = \omega^2 \epsilon \cdot \mathbb{A}_j + \sigma \cdot (-i\omega \mathbb{A}_j)$$

$$(i\omega\sigma - \omega^2 \epsilon) \cdot \mathbb{A}_j + \nabla \times \mu^{-1} \nabla \times \mathbb{A}_j = 0 \tag{A.426}$$

Appendix B. Material properties and constants

Material	electrical properties			thermal properties			mechanical properties		
Pt	$\varrho_0 =$	$10.6 \cdot 10^{-8}$ Ωm	[106]	$\kappa =$	71.6 W/mK	[106]	$E =$	$1.65 \cdot 10^{11}$ N/m^2	[cf. 56]
	$\alpha =$	$3 \cdot 10^{-3}$ 1/K	*	$\rho =$	21450 kg/m^3	[106]	$\nu =$	0.39	[62]
	$\mu_r =$	$1 + 2.57 \cdot 10^{-4} -$	[121]	$c_p =$	130 J/kgK	[cf. 106]	$\beta =$	$9 \cdot 10^{-6}$	[92]
STO				$\kappa =$	10.305 W/mK	*	$E =$	$2.67 \cdot 10^{11}$ N/m^2	[164]
SrTiO$_3$				$\rho =$	5120 kg/m^3	[45]	$\nu =$	0.24	[cf. 41]
				$c_p =$	544 J/kgK	[51]	$\beta =$	$9.4 \cdot 10^{-6}$	[cf. 149]
PCMO	$R(T) =$	$77.86/\sqrt{T} \cdot e^{1272.26/T}\,\Omega$	*	$\kappa =$	1.58 W/mK	[43]	$E =$	$3.00 \cdot 10^{11}$ N/m^2	
Pr$_{1-x}$Ca$_x$MnO$_3$				$\rho =$	570 kg/m^3	[cf. 160]	$\nu =$	0.4	
				$c_p =$	480 J/kgK	[cf. 102]	$\beta =$	$10 \cdot 10^{-6}$	
MgO				$\kappa =$	56.975 W/mK	*	$E =$	$3.16 \cdot 10^{11}$ N/m^2	[116]
				$\rho =$	3585 kg/m^3	[94]	$\nu =$	0.18	[116]
				$c_p =$	873.6 J/kgK	[94]			
Cu				$\kappa =$	400 W/mK	[cf. 109]			
				$\rho =$	8960 kg/m^3	[106]			
				$c_p =$	385 J/kgK	[106]			
YSZ				$\kappa =$	1.8 W/mK	[147]	$E =$	$2.20 \cdot 10^{11}$ N/m^2	[cf. 97]
ZrO$_2$/Y$_2$O$_3$				$\rho =$	5900 kg/m^3	[147]	$\nu =$	0.25	[162]
				$c_p =$	480 J/kgK	[147]	$\beta =$	$9.2 \cdot 10^{-6}$	[cf. 141]
Si				$\kappa =$	149 W/mK	[136]	$E =$	$1.30 \cdot 10^{11}$ N/m^2	[112]
				$\rho =$	2330 kg/m^3	[106, 171]	$\nu =$	0.28	[112]
				$c_p =$	700 J/kgK	[15, 171]	$\beta =$	$2.6 \cdot 10^{-6}$	[149]
SIN				$\kappa =$	2.4 W/mK	**	$E =$	$2.9 \cdot 10^{11}$ N/m^2	[9]
Si$_3$N$_4$				$\rho =$	3170 kg/m^3	[106]	$\nu =$	0.22	[9]
				$c_p =$	710 J/kgK	[114, 166]	$\beta =$	$2.3 \cdot 10^{-6}$	[42]
PMMA				$\kappa =$	0.2 W/mK	[34]			
C$_5$H$_8$O$_2$				$\rho =$	1200 kg/m^3	[123]			
				$c_p =$	1470 J/kgK	[cf. 4, 84]			
W				$\kappa =$	174 W/mK	[106]			
				$\rho =$	19300 kg/m^3	[106]			
				$c_p =$	132 J/kgK	[106]			

Table 5: Collection of all used material parameters. Values indicated with * are determined within the joint project. The value indicated with ** is by courtesy of F. Völklein, Institute for Microtechnologies, RheinMain University of Applied Sciences Wiesbaden, Germany.

Name		Value	
Stefan-Boltzmann constant	$\sigma_b =$	$5.670 \cdot 10^{-8}$ W/m^{2}K^4	[106]
electric permittivity vacuum constant	$\epsilon_0 =$	$8.854187 \cdot 10^{-12}$ F/m	[106]
magnetic permeability vacuum constant	$\mu_0 =$	$4\pi \cdot 10^{-7}$ N/A^2	[106]
Euler-Maschoneri constant	$\gamma =$	0.577215...	[37]

Table 6: Collection of used constants.

Appendix C. Technical drawing of the 6-pad heater structure

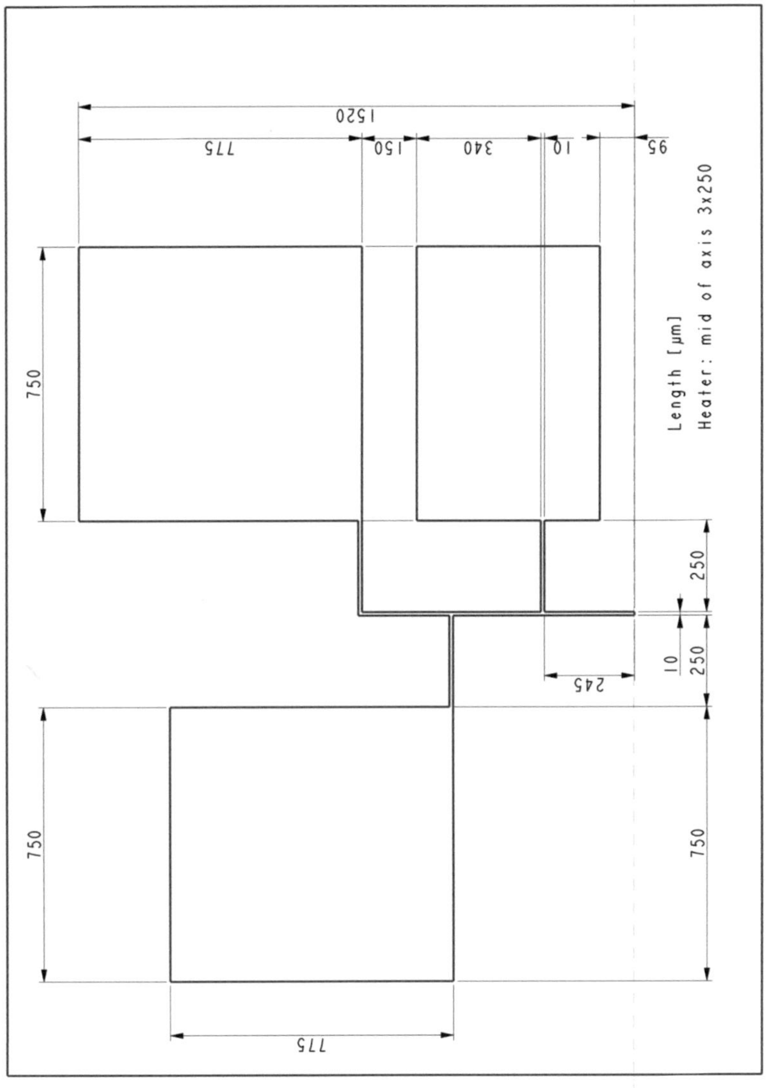

Figure 116: Technical drawing of the 6-pad heater structure.

Publications

S. WIEDIGEN, M. FEUCHTER, CH. JOOSS AND M. KAMLAH, *3-ω measurements of thermal conductivity in oxide thin films*, poster at the 'DPG-conference' in Regensburg, Germany (2010)

S. WIEDIGEN, M. FEUCHTER, M. KAMLAH AND CH. JOOSS, *3ω measurements of thermal conductivity in oxide thin films*, oral presentation at the '3 Omega workshop' of the priority program in Göttingen, Germany (2010)

M. FEUCHTER AND M. KAMLAH, *Numerical analysis of the 3-ω-method for measuring thermal conductivities in thermoelectric materials at the nanoscale*, poster at the 'HEAT conference' in Erice, Italy (2010)

S. WIEDIGEN, M. FEUCHTER, CH. JOOSS AND M. KAMLAH, *Understanding size and interface dependent anisotropic thermal conduction in correlated multilayer structures*, poster at the 'status meeting of the priority program' in Berlin, Germany (2010)

S. WIEDIGEN, M. FEUCHTER, K. R. MANGIPUDI, CH. JOOSS, M. KAMLAH, C. VOLKERT, *Thermal conduction in correlated multilayer structures*, poster at the 'DPG-conference' in Dresden, Germany (2011)

S. WIEDIGEN, T. KRAMER, K. R. MANGIPUDI, M. FEUCHTER, CH. JOOSS, C. VOLKERT AND M. KAMLAH, *Numerical analysis of the 3ω-method for measuring thermal conductivities in thermoelectric materials at the nanoscale*, poster at the 'status meeting of the priority program' in Wittenberg, Germany (2011)

S. Wiedigen, T. Kramer, N. Folchert, J. Norpoth, Ch. Jooss, M. Feuchter, M. Kamlah, K. R. Mangipudi, C. A. Volkert, *Thermal conductivity in correlated oxides*, poster at the 'status meeting of the priority program' in Wittenberg, Germany (2011)

M. Feuchter and M. Kamlah, *Inverse Geometry configurations for the 3ω-method*, oral presentation at the 'priority program summer school' in Cuxhaven, Germany (2011)

M. Feuchter and M. Kamlah, *Numerical analysis of the 3ω-method*, oral presentation at the 'COMSOL conference' in Stuttgart, Germany (2011)

M. Feuchter and M. Kamlah, *Numerical analysis of the 3ω-method*, oral presentation at the 'Messtechnik workshop' of the priority program in Kaub am Rhein, Germany (2012)

S. Wiedigen, T. Kramer, M. Feuchter K. R. Mangipudi, J. Hoffmann, M. Kamlah, C. A. Volkert and Ch. Jooss, *Thermal conductivity in $Pr_{1-x}Ca_xMnO_3$ and $SrTiO_3$ thin film systems*, poster at the 'Messtechnik workshop' of the priority program in Kaub am Rhein, Germany (2012)

S. Wiedigen, T. Kramer, M. Feuchter, I. Knorr, N. Nee, J. Hoffmann, M. Kamlah, C. A. Volkert and Ch. Jooss, *Interplay of point defects, biaxial strain, and thermal conductivity in homoepitaxial $SrTiO_3$ thin films*, Applied Physics Letters, 100 (2012), pp. 061904-4

M. Feuchter, S. Wiedigen, Ch. Jooss and M. Kamlah, *A method to determine the anisotropic thermal conduction of nanoscale thermoelectric materials by the use of multiphysical simulations and neural networks*, oral presentation at the 'International Conference on Computational Methods ICCM' in Gold Coast, Australia (2012)

M. Feuchter and M. Kamlah, *Simulation of heat transfer in a 3ω-measurement*, oral presentation at the 'Thermal Transport at the Nanoscale' workshop of the priority program in Bad Honnef, Germany (2013)

M. Feuchter and M. Kamlah, *Simulation of heat transfer in a 3ω-measurement*, oral presentation at the 'Germany/Japan Seminar on Advanced Materials for Zero-Emission Energy' workshop in Kyoto, Japan (2014)

References

[1] M. Abramowitz and I. A. Stegun, *Handbook of mathematical functions with formulas, graphs, and mathematical tables*, U.S. Dept. of Commerce : U.S. G.P.O., Washington, D.C., (1972).

[2] S. Ahmed, R. Liske, T. Wunderer, M. Leonhardt, R. Ziervogel, C. Fansler, T. Grotjohn, J. Asmussen, and T. Schuelke, *Extending the 3ω-method to the MHz range for thermal conductivity measurements of diamond thin films*, Diamond and Related Materials, 15 (2006), pp. 389–393.

[3] J. Alvarez-Quintana and J. Rodraguez-Viejo, *Extension of the 3ω method to measure the thermal conductivity of thin films without a reference sample*, Sensors and Actuators A: Physical, 142 (2008), pp. 232–236.

[4] S. P. Andersson and R. G. Ross, *Thermal Conductivity and Heat Capacity per Unit Volume of Poly(methyl Methacrylate) Under High Pressure*, International Journal of Thermophysics, 15, 5 (1994), pp. 949–962.

[5] G. Antonini, *Internal impedance of conductors of rectangular cross section*, IEEE Transactions on Microwave Theory and Techniques, 47, 7 (1999), pp. 979–985.

[6] ASM International, *Materials and Coatings for Medical Devices: Cardiovascular*, ASM International, Materials Park, Ohio, (2009).

[7] A. Baikalov, Y. Q. Wang, B. Shen, B. Lorenz, S. Tsui, Y. Y. Sun, Y. Y. Xue, and C. W. Chu, *Field-driven hysteretic and reversible resistive switch at the Ag-Pr 0.7 Ca 0.3 MnO 3 interface*, Applied Physics Letters, 83 (2003), pp. 957–959.

[8] H. Balke, *Einführung in die Technische Mechanik Festigkeitslehre*, Springer-Verlag Berlin Heidelberg, Berlin, Heidelberg, (2010).

[9] H. Baltes, O. Brand, G. K. Fedder, C. Hierold, J. Korvink, and O. Tabata, *CMOS - MEMS; Advanced Micro & Nanosystems Volume 2*, Wiley-VCH, (2005).

[10] G. M. Barrow and G. W. Herzog, *Physikalische Chemie : Gesamtausgabe*, Bohmann, Wien, (1984).

[11] H. Bateman, *Tables of integral transforms Volume I.*, McGraw-Hill Book Company INC., New York, N.Y., (1954).

[12] J.-L. Battaglia, C. Wiemer, and M. Fanciulli, *An accurate low-frequency model for the 3ω method*, Journal of Applied Physics, 101 (2007), pp. 104510–5.

[13] T. Bergman, F. Incropera, A. Lavine, and D. DeWitt, *Fundamentals of Heat and Mass Transfer*, Wiley, (2011).

[14] R. P. Bhatta, S. Annamalai, R. K. Mohr, M. Brandys, I. L. Pegg, and B. Dutta, *High temperature thermal conductivity of platinum microwire by 3ω method*, Review of Scientific Instruments, 81 (2010), pp. 114904–5.

[15] H. H. Binder, *Lexikon der chemischen Elemente : das Periodensystem in Fakten, Zahlen und Daten*, S. Hirzel, Stuttgart, (1999).

[16] N. O. Birge, *Specific-heat spectroscopy of glycerol and propylene glycol near the glass transition*, Physical Review B, 34 (1986), pp. 1631–1642.

[17] N. O. Birge and S. R. Nagel, *Specific-heat spectroscopy of the glass transition*, Physical Review Letters, 54 (1985), pp. 2674–2677.

[18] O. Biro, P. Böhm, K. Preis, and G. Wachutka, *Edge Finite Element Analysis of Transient Skin Effect Problems*, IEEE Transactions on Magnetics, 36, 4 (2000), pp. 835–839.

[19] O. Biro and K. Preis, *On the Use of the Magnetic Vector Votential in the Finite Element Analysis of Three-Dimensional Eddy Currents*, IEEE Transactions on Magnetics, 25 (1989), pp. 3145–3159.

[20] O. Biro, K. Preis, G. Buchgraber, and I. Ticar, *Voltage-Driven Coils in Finite-Element Formulations Using a Current Vector and a Magnetic Scalar Potential*, IEEE Transactions on Magnetics, 40, 2 (2004), pp. 1286–1289.

[21] P. Böckh, *Wärmeübertragung Grundlagen und Praxis*, Springer-Verlag Berlin Heidelberg, Berlin, Heidelberg, (2006).

[22] T. Borca-Tasciuc, A. R. Kumar, and G. Chen, *Data reduction in 3ω method for thin-film thermal conductivity determination*, Review of Scientific Instruments, 72 (2001), pp. 2139–2147.

[23] T. Borca-Tasciuc, D. W. Song, J. R. Meyer, I. Vurgaftman, M.-J. Yang, B. Z. Nosho, L. J. Whitman, H. Lee, R. U. Martinelli, G. W. Turner, M. J. Manfra, and G. Chen, *Thermal conductivity of alas0.07sb0.93 and al0.9ga0.1as0.07sb0.93 alloys and (alas)1/(alsb)11 digital-alloy superlattices*, Journal of Applied Physics, 92 (2002), pp. 4994–4998.

[24] T. Borca-Tasciuc, S. Vafaei, D.-A. Borca-Tasciuc, B. Q. Wei, R. Vajtai, and P. M. Ajayan, *Anisotropic thermal diffusivity of*

aligned multiwall carbon nanotube arrays, Journal of Applied Physics, 98 (2005), pp. 054309–6.

[25] A. I. BOUKAI, Y. BUNIMOVICH, J. TAHIR-KHELI, J.-K. YU, W. A. GODDARD III, AND J. R. HEATH, *Silicon nanowires as efficient thermoelectric materials*, Nature, 451 (2008), pp. 168–171.

[26] R. N. BRACEWELL, *The Fourier Transform and Its Applications*, McGraw-Hill Book Company INC., (2000).

[27] I. N. BRONSTEIN, K. A. SEMENDJAJEW, G. MUSIOL, AND H. MÜHLIG, *Taschenbuch der Mathematik*, Verlag Harri Deutsch, (2005).

[28] D. G. CAHILL, *Thermal conductivity measurement from 30 to 750 K: the 3ω method*, Review of Scientific Instruments, 61 (1990), pp. 802–808.

[29] D. G. CAHILL, *Erratum: "Thermal conductivity measurement from 30 to 750 K: the 3ω method, Review of Scientific Instruments, 61 (1990), pp. 802-808"*, Review of Scientific Instruments, 73 (2002), pp. 3701–3701.

[30] D. G. CAHILL AND T. H. ALLEN, *Thermal conductivity of sputtered and evaporated SiO2 and TiO2 optical coatings*, Applied Physics Letters, 65 (1994), pp. 309–311.

[31] D. G. CAHILL, W. K. FORD, K. E. GOODSON, G. D. MAHAN, A. MAJUMDAR, H. J. MARIS, R. MERLIN, AND S. R. PHILLPOT, *Nanoscale thermal transport*, Journal of Applied Physics, 93 (2003), pp. 793–818.

[32] D. G. CAHILL, M. KATIYAR, AND J. R. ABELSON, *Thermal conductivity of a-Si:H thin films*, Physical Review B, 50 (1994), pp. 6077–6081.

[33] D. G. CAHILL, S.-M. LEE, AND T. I. SELINDER, *Thermal conductivity of κ-Al2O3 and α-Al2O3 wear-resistant coatings*, Journal of Applied Physics, 83 (1998), pp. 5783–5786.

[34] D. G. CAHILL AND R. O. POHL, *Thermal conductivity of amorphous solids above the plateau*, Physical Review B, 35 (1987), pp. 4067–4073.

[35] H. S. CARSLAW AND J. C. JAEGER, *Conduction of heat in solids, Second Edition*, Clarendon Press - Oxford, (1959).

[36] W.-Y. CHANG, J.-H. LIAO, Y.-S. LO, AND T.-B. WU, *Resistive switching characteristics in Pr0.7Ca0.3MnO3 thin films on LaNiO3-electrodized Si substrate*, Applied Physics Letters, 94 (2009), pp. 172107–3.

[37] C.-P. CHEN AND C. MORTICI, *New sequence converging towards the Euler-Mascheroni constant*, Computers and Mathematics with Applications, 64 (2012), pp. 391–398.

[38] F. CHEN, J. SHULMAN, Y. XUE, C. W. CHU, AND G. S. NOLAS, *Thermal conductivity measurement under hydrostatic pressure using the 3ω method*, Review of Scientific Instruments, 75 (2004), pp. 4578–4584.

[39] Y. CHEN, W. TIAN, H. LI, X. WANG, AND W. ZHU, *PCMO Device With High Switching Stability*, IEEE Electron Device Letters, 31 (2010), pp. 866–868.

[40] S. R. CHOI, J. KIM, AND D. KIM, *3ω method to measure thermal properties of electrically conducting small-volume liquid*, Review of Scientific Instruments, 78 (2007), pp. 084902–6.

[41] H. M. Christen, E. D. Specht, S. S. Silliman, and K. S. Harshavardhan, *Ferroelectric and antiferroelectric coupling in superlattices of paraelectric perovskites at room temperature*, Physical Review B, 68 (2003), p. 020101.

[42] W.-H. Chuang, T. Luger, R. K. Fettig, and R. Ghodssi, *Mechanical Property Characterization of LPCVD Silicon Nitride Thin Films at Cryogenic Temperatures*, Journal of Microelectromechanical Systems, 13 (5) (2004), pp. 870–879.

[43] B. T. Cong, T. Tsuji, P. X. Thao, P. Q. Thanh, and Y. Yamamura, *High-temperature thermoelectric properties of $Ca_{1-x}Pr_xMnO_{3-\delta}(0 \leq x < 1)$*, Physica B: Condensed Matter, 352 (2004), pp. 18–23.

[44] O. Corbino, *Periodische Widerstandsänderungen von Metallfäden.*, Physikalische Zeitschrift, 12 (1911), pp. 292–295.

[45] crystec. *http://www.crystec.de/ daten/ srtio3.pdf*, (03:20 p.m.,14 Jun. 2013).

[46] P. Csíkvári, P.Fürjes, C. Dücső, and I. Bársony, *Micro-hotplates for thermal characterisation of structural materials of MEMS*, Microelectronics Journal, 40 (2009), pp. 1393–1397.

[47] C. Dames and G. Chen, *1ω, 2ω, and 3ω methods for measurements of thermal properties*, Review of Scientific Instruments, 76 (2005), pp. 124902–14.

[48] A. Daoud-Aladine, J. Rodrguez-Carvajal, L. Pinsard-Gaudart, M. T. Fernndez-Daz, and A. Revcolevschi, *Zener Polaron Ordering in Half-Doped Manganites*, Physical Review Letters, 89 (2002), p. 097205.

[49] N. Das, S. Tsui, Y. Y. Xue, Y. Q. Wang, and C. W. Chu, *Electric-field-induced submicrosecond resistive switching*, Physical Review B, 78 (2008), pp. 235418–.

[50] R. M. Dreizler and C. S. Lüdde, *Theoretische Physik 2 - Elektrodynamik und spezielle Relativitätstheorie*, Springer-Verlag Berlin Heidelberg, Berlin, Heidelberg, (2005).

[51] A. Durán, F. Morales, L. Fuentes, and J. M. Siqueiros, *Specific heat anomalies at 37, 105 and 455 K in $SrTiO_3$: Pr*, Journal of Physics: Condensed Matter, 20 (2008), p. 085219.

[52] S. N. Dyer, *Survey of instrumentation and measurement*, Wiley-Interscience, John Wiley and Sons, Inc., New York, (2001).

[53] D. V. Efremov, J. van den Brink, and D. I. Khomskii, *Charge order versus Zener polarons: ferroelectricity in manganites*, Condensed Matter, 1 (2003), p. 0306651.

[54] D. V. Efremov, J. van den Brink, and D. I. Khomskii, *Bond-versus site-centred ordering and possible ferroelectricity in manganites*, Nature Materials, 3 (2004), pp. 853–856.

[55] electric ray. *http://electric-ray.com/ Tesla_ Roadster_ cutaway.jpg*, (10:06 a.m., 23 Jan. 2013).

[56] R. Farraro and R. McLellan, *Temperature Dependence of the Young's Modulus and Shear Modulus of Pure Nickel, Platinum, and Molybdenum*, Metallurgical Transactions A, 8 (1977), pp. 1563–1565.

[57] R. P. Feynman, *Lectures on Physics*, Addison-Wesley Pub. Co., (1963).

[58] H. Ftouni, C. Blanc, A. Sikora, J. Richard, M. Defoort, K. Lulla, E. Collin, and O. Bourgeois, *Thermal conductivity measurement of suspended Si-N membranes from 10 K to 275 K using the 3ω-Völklein method*, Journal of Physics: Conference Series, 395 (2012), p. 012109.

[59] M. Fujimoto, H. Koyama, Y. Nishi, and T. Suzuki, *Resistive switching properties of high crystallinity and low-resistance Pr0.7Ca0.3MnO3 thin film with point-contacted Ag electrodes*, Applied Physics Letters, 91 (2007).

[60] R. Gasch, K. Knothe, and R. Liebich, *Strukturdynamik - Diskrete Systeme und Kontinua*, Springer-Verlag Berlin Heidelberg, (2012).

[61] A. K. Gasiorski, *A finite element method for analysing skin effect in conductors with unknown values of surface boundary conditions*, International Journal for Numerical Methods in Engineering, 21 (1985), pp. 1641–1657.

[62] H. Gercek, *Poisson's ratio values for rocks*, International Journal of Rock Mechanics and Mining Sciences, 44 (2007), pp. 1–13.

[63] C. Gerthsen, H. O. Kneser, and H. Vogel, *Physik*, Springer-Verlag Berlin Heidelberg New York, (1974).

[64] M. Greconici, G. Madescu, and M. Mot, *Skin Effect Analysis in a Free Space Conductor*, SER.: ELEC. ENERG., 23, 2 (2010), pp. 207–215.

[65] greencar. *http://cdn-www.greencar.com/ images/ waste- exhaust- heat- generates- electricity- cars- efficient.php/ bmw-teg-1.jpg*, (10:08 a.m., 23 Jan. 2013).

[66] D. J. Griffiths, *Introduction to Electrodynamics*, Prentice-Hall, Upper Saddle River, N.J., (1999).

[67] D. Gross, W. Hauger, and P. Wriggers, *Technische Mechanik 4*, Springer, Berlin; Heidelberg, (2009).

[68] D. Gross and T. Seelig, *Bruchmechanik - Mit einer Einführung in die Mikromechanik*, Springer-Verlag, Berlin; New York, (2006).

[69] S. P. Gurrum, W. P. King, and Y. K. Joshi, *A semianalytical solution for the 3ω method including the effect of heater thermal conduction*, Journal of Applied Physics, 103 (2008), pp. 113517–6.

[70] L. D. Hicks and M. S. Dresselhaus, *Effect of quantum-well structures on the thermoelectric figure of merit*, Physical Review B, 47 (1993), pp. 12727–12731.

[71] C. Y. Ho, R. W. Powell, and P. E. Liley, *Thermal Conductivity of the Elements*, Journal of Physical and Chemical Reference Data, 1 (1972), p. 279.

[72] E. A. Hoffmann, H. A. Nilsson, J. E. Matthews, N. Nakpathomkun, A. I. Persson, L. Samuelson, and H. Linke, *Measuring Temperature Gradients over Nanometer Length Scales*, Nano Letters, 9 (2009), pp. 779–783.

[73] N. Huber, *Anwendung Neuronaler Netze bei nichtlinearen Problemen der Mechnaik*, FZKA 6504, (2000).

[74] R. E. Hummel, *Electronic Properties of Materials*, Springer, New York, (2005).

[75] INTERNATIONAL ENERGY AGENCY, *World energy outlook 2012 - executive summary*, http://www.worldenergyoutlook.org/, (2012).

[76] J. D. JACKSON, *Klassische Elektrodynamik*, Walter de Gruyter, Berlin, (1982).

[77] A. JACQUOT, G. CHEN, H. SCHERRER, A. DAUSCHER, AND B. LENOIR, *Improvements of on-membrane method for thin film thermal conductivity and emissivity measurements*, Sensors and Actuators A: Physical, 117 (2005), pp. 203–210.

[78] A. JACQUOT, B. LENOIR, A. DAUSCHER, M. STOLZER, AND J. MEUSEL, *Numerical simulation of the 3ω method for measuring the thermal conductivity*, Journal of Applied Physics, 91 (2002), pp. 4733–4738.

[79] A. JACQUOT, F. VOLLMER, B. BAYER, M. JAEGLE, D. EBLING, AND H. BOETTNER, *Thermal Conductivity Measurements on Challenging Samples by the 3 Omega Method*, Journal of Electronic Materials, 39 (2010), pp. 1621–1626.

[80] T. JAEGER, *Thermoelectric properties of TiNiSn and $Zr_{0.5}Hf_{0.5}NiSn$ thin films and superlattices with reduced thermal conductivities*, Ph.D.-thesis, Johannes Gutenberg-Universität, Mainz, Germany, (2013).

[81] T. JAEGER, C. MIX, M. SCHWALL, X. KOZINA, J. BARTH, B. BALKE, M. FINSTERBUSCH, Y. U. IDZERDA, C. FELSER, AND G. JAKOB, *Epitaxial growth and thermoelectric properties of TiNiSn and $Zr_{0.5}Hf_{0.5}NiSn$ thin films*, Thin Solid Films, 520 (2011), pp. 1010–1014.

[82] A. JAIN AND K. E. GOODSON, *Measurement of the Thermal Conductivity and Heat Capacity of Freestanding Shape Memory Thin Films Using the 3ω Method*, Journal of Heat Transfer, 130 (2008), pp. 102402–7.

[83] K. JAKUBIUK AND P. ZIMNY, *Skin effect in rectangular conductors*, Journal of Physics A: Mathematical and General, 9, 4 (1976), pp. 669–676.

[84] R. JENISCH AND M. STOHRER, *Wärme*, in Lehrbuch der Bauphysik, Vieweg+Teubner, (2008), pp. 107–335.

[85] C. JOOSS, L.WU, T. BEETZ, R. F. KLIE, M. BELEGGIA, M. A. SCHOFIELD, S. SCHRAMM, J. HOFFMANN, AND Y. ZHU, *Polaron melting and ordering as key mechanisms for colossal resistance effects in manganites*, Proceedings of the National Academy of Science, 104 (2007), pp. 13597–13602.

[86] M. KAMLAH, *Zur Modellierung des Verfestigungsverhaltens von Materialien mit statischer Hysterese im Rahmen der phänomenologischen Thermomechanik*, PhD thesis, Universität Gesamthochschule Kassel, (1994).

[87] R. KARTHIK, R. HARISH NAGARAJAN, B. RAJA, AND P. DAMODHARAN, *Thermal conductivity of CuO-DI water nanofluids using 3-ω measurement technique in a suspended micro-wire*, Experimental Thermal and Fluid Science, xx (2012), pp. –.

[88] S. P. KEANE, S. SCHMIDT, J. LU, A. E. ROMANOV, AND S. STEMMER, *Phase transitions in textured SrTiO₃ thin films on epitaxial Pt electrodes*, Journal of Applied Physics, 99 (2006), p. 033521.

[89] C. J. Kim and I.-W. Chen, *Effect of top electrode on resistance switching of (Pr,Ca)MnO₃ thin films*, Thin Solid Films, 515 (2006), pp. 2726–2729.

[90] D. S. Kim, Y. H. Kim, C. E. Lee, and Y. T. Kim, *Colossal electroresistance mechanism in a Au/Pr$_{0.7}$Ca$_{0.3}$MnO$_3$/pt sandwich structure: Evidence for a Mott transition*, Physical Review B, 74 (2006), p. 174430.

[91] J. H. Kim, A. Feldman, and D. Novotny, *Application of the three omega thermal conductivity measurement method to a film on a substrate of finite thickness*, Journal of Applied Physics, 86 (1999), pp. 3959–3963.

[92] R. K. Kirby, *Platinum - A Thermal Expansion Reference Material*, International Journal of Thermophysics, 12 , 4 (1991).

[93] K. Knothe and H. Wessels, *Finite Elemente - Eine Einführung für Ingenieure*, Springer-Verlag Berlin Heidelberg, (2008).

[94] korth. *http://www.korth.de/ index.php/ material- detailansicht/ items/22.html*, (05.04 p.m.,14. Jun. 2013).

[95] K. Koumoto, Y. Wang, R. Zhang, A. Kosuga, and R. Funahashi, *Oxide Thermoelectric Materials: A Nanostructuring Approach*, Annual Review of Materials Research, 40 (2010), pp. 363–394.

[96] R. Kruse, *Computational Intelligence : Eine methodische Einführung in Künstliche Neuronale Netze, Evolutionäre Algorithmen, Fuzzy-Systeme und Bayes-Netze; 43 Tab.*, Vieweg + Teubner, Wiesbaden, (2011).

[97] K. Kurosaki, D. Setoyama, J. Matsunaga, and S. Yamanaka, *Nanoindentation tests for TiO_2, MgO, and YSZ single crystals*, Journal of Alloys and Compounds, 386 (2005), pp. 261–264.

[98] M. Leclerc and A. Najari, *Organic thermoelectrics: Green energy from a blue polymer*, Nature Materials, 10 (2011), pp. 409–410.

[99] S.-M. Lee and D. G. Cahill, *Heat transport in thin dielectric films*, Journal of Applied Physics, 81 (1997), pp. 2590–2595.

[100] S.-M. Lee, D. G. Cahill, and T. H. Allen, *Thermal conductivity of sputtered oxide films*, Physical Review B, 52 (1995), pp. 253–257.

[101] S.-M. Lee, D. G. Cahill, and R. Venkatasubramanian, *Thermal conductivity of Si-Ge superlattices*, Applied Physics Letters, 70 (1997), pp. 2957–2959.

[102] M. R. Lees, O. A. Petrenko, G. Balakrishnan, and D. McK. Paul, *Specific heat of $Pr_{0.6}(Ca_{1-x}Sr_x)_{0.4}MnO_3$ ($0 \leq x \leq 1$)*, Physical Review B, 59 (2) (1999), pp. 1298–1303.

[103] P. Leonard and D. Rodger, *Finite element scheme for transient 3D eddy currents*, IEEE Transactions on Magnetics, 24 (1988), pp. 90–93.

[104] S.-L. Li, D. S. Shang, J. Li, J. L. Gang, and D. N. Zheng, *Resistive switching properties in oxygen-deficient Pr0.7Ca0.3MnO3 junctions with active Al top electrodes*, Journal of Applied Physics, 105 (2009).

[105] Z. L. Liao, Z. Z. Wang, Y. Meng, Z. Y. Liu, P. Gao, J. L. Gang, H. W. Zhao, X. J. Liang, X. D. Bai, and D. M. Chen, *Categorization of resistive switching of*

metal-Pr0.7Ca0.3MnO3-metal devices, Applied Physics Letters, 94 (2009).

[106] D. R. Lide, *CRC Handbook of Chemistry and Physics*, CRC Press, Boca Raton, Florida, Boca Raton, Fla., (2009).

[107] M. E. Lines and A. M. Glass, *Principles and Applications of Ferroelectrics and Related Materials*, Clarendon Press - Oxford, Oxford, (1977).

[108] S. Q. Liu, N. J. Wu, and A. Ignatiev, *Electric-pulse-induced reversible resistance change effect in magnetoresistive films*, Applied Physics Letters, 76 (2000), pp. 2749–2751.

[109] W. Liu, Y. Yang, and M. Asheghi, *Thermal and electrical characterization and modeling of thin copper layers*, ITHERM '06. The Tenth Intersociety Conference, (2006), pp. 1171–1176.

[110] L. Lu, W. Yi, and D. L. Zhang, *3ω method for specific heat and thermal conductivity measurements*, Review of Scientific Instruments, 72 (2001), pp. 2996–3003.

[111] J. Lunze, *Regelungstechnik 1*, Springer-Verlag Berlin Heidelberg, (2007).

[112] H. M.A., N. W.D., and K. T.W., *What is the Young's Modulus of Silicon?*, Journal of Microelectromechanical Systems, 19, 2 (2010), pp. 229–238.

[113] L. E. Malvern, *Introduction to the mechanics of a continuous medium*, Prentice-Hall, Inc., Englewood Cliffs, N.J., (1969).

[114] C. H. Mastrangelo, Y.-C. Tai, and R. S. Muller, *Thermophysical properties of low-residual stress, Silicon-rich, LPCVD silicon nitride films*, Sensors and Actuators A: Physical, 23 (1990), pp. 856–860.

[115] J. C. Maxwell, *A Dynamical Theory of the Electromagnetic Field*, Philosophical Transactions of the Royal Society of London, 155 (1865), pp. 459–512.

[116] T. E. Mitchell, *Dislocations and Mechanical Properties of MgO-Al$_2$O$_3$ Spinel Single Crystals*, Journal of the American Ceramic Society, 82 (12) (1999), pp. 3305–3316.

[117] I. K. Moon, Y. H. Jeong, and S. I. Kwun, *The 3ω technique for measuring dynamic specific heat and thermal conductivity of a liquid or solid*, Review of Scientific Instruments, 67 (1996), pp. 29–35.

[118] R. E. Moritz, *Memorabilia mathematica; or, The philomath's quotation-book.*, The Macmillan Company, (1914).

[119] n tv. *http://bilder3.n-tv.de/ img/ incoming/ origs6923101/ 3272734610- w1000- h960/ curiosity2.jpg*, (09:45 a.m., 23 Jan. 2013).

[120] C. Nauenheim, C. Kügeler, A. Rüdiger, R. Waser, A. Flocke, and T. Noll, *Nano-Crossbar Arrays for Nonvolatile Resistive RAM (RRAM) Applications*, NANO'08. 8th IEEE conference on Nanotechnology, (2008), pp. 464–467.

[121] ndtguide. *http://ndtguide.com/ wiki/ mp/ Permeability.pdf*, (10:43 a.m., 14 Jun. 2013).

[122] K. Nielsch, J. Bachmann, J. Kimling, and H. Bttner, *Thermoelectric Nanostructures: From Physical Model Systems*

towards Nanograined Composites, Advanced Energy Materials, 1 (2011), pp. 713–731.

[123] Y. NOMURA, R. HORLEIN, P. TZALLAS, B. DROMEY, S. RYKOVANOV, Z. MAJOR, J. OSTERHOFF, S. KARSCH, L. VEISZ, M. ZEPF, D. CHARALAMBIDIS, F. KRAUSZ, AND G. D. TSAKIRIS, *Attosecond phase locking of harmonics emitted from laser-produced plasmas*, Nature Physics, 5 (2009), pp. 124–128.

[124] D.-W. OH, A. JAIN, J. K. EATON, K. E. GOODSON, AND J. S. LEE, *Thermal conductivity measurement and sedimentation detection of aluminum oxide nanofluids by using the 3ω method*, International Journal of Heat and Fluid Flow, 29 (2008), pp. 1456–1461.

[125] H. OHTA, S. KIM, Y. MUNE, T. MIZOGUCHI, K. NOMURA, S. OHTA, T. NOMURA, Y. NAKANISHI, Y. IKUHARA, M. HIRANO, H. HOSONO, AND K. KOUMOTO, *Giant thermoelectric Seebeck coefficient of a two-dimensional electron gas in SrTiO₃*, Nature Materials, 6 (2007), pp. 129–134.

[126] B. W. OLSON, S. GRAHAM, AND K. CHEN, *A practical extension of the 3ω method to multilayer structures*, Review of Scientific Instruments, 76 (2005), pp. 053901–7.

[127] B. K. PARK, J. PARK, AND D. KIM, *Note: Three-omega method to measure thermal properties of subnanoliter liquid samples*, Review of Scientific Instruments, 81 (2010), pp. 066104–3.

[128] M. PAZ AND W. LEIGH, *Structural Dynamics - Theory and Computation*, Kluwer Academic, (2004).

[129] P. PICHANUSAKORN AND P. BANDARU, *Nanostructured thermoelectrics*, Materials Science and Engineering: R: Reports, 67 (2010), pp. 19–63.

[130] Z. B. POPOVIC AND B. D. POPOVIC, *Introductory Electromagnetics*, Addison-Wesley Pub. Co., (1999).

[131] B. POUDEL, Q. HAO, Y. MA, Y. LAN, A. MINNICH, B. YU, X. YAN, D. WANG, A. MUTO, D. VASHAEE, X. CHEN, J. LIU, M. S. DRESSELHAUS, G. CHEN, AND Z. REN, *High-Thermoelectric Performance of Nanostructured Bismuth Antimony Telluride Bulk Alloys*, Science, 320 (2008), pp. 634–638.

[132] S. S. RAO, *The Finite Element Method in Engineering*, Butterworth-Heinemann, imprint of Elsevier, (2011).

[133] C. E. RAUDZIS, *Anwendung und Erweiterung der 3ω Methode zur Charakterisierung komplexer Mehrschichtsysteme in der Mikrosystemtechnik*, Der Andere Verlag, Tönning Lübeck Marburg, (2007).

[134] C. E. RAUDZIS, F. SCHATZ, AND D. WHARAM, *Extending the 3ω method for thin-film analysis to high frequencies*, Journal of Applied Physics, 93 (2003), pp. 6050–6055.

[135] RAUMFAHRER. *http://www.raumfahrer.net/ news/ images/ orion_ esa_ sm1.jpg*, (10:26 a.m., 23 Jan. 2013).

[136] S.-C. ROH, Y.-M. CHOI, AND S.-Y. KIM, *Sensitivity enhancement of a silicon micro-machined thermal flow sensor*, Sensors and Actuators A: Physical, 128 (2006), pp. 1–6.

[137] D. ROWE, *CRC Handbook of Thermoelectrics*, CRC Press, Boca Raton, (1995).

[138] D. M. Rowe and C. M. Bhandari, *Modern Thermoelectrics*, Reston Publishing Company, Inc., (1983).

[139] A. Salvatore, *Finite-Element Mesh Adaptation of 2-D Time-Harmonic Skin Effect Problems*, IEEE Transactions on Magnetics, 36, 4 (2000), pp. 1592–1595.

[140] S. Scherrer-Rudiy, *Bewertung von Unsicherheiten bei werkstoffmechanischen Fragestellungen mit dem Bayes-Verfahren*, PhD thesis, Forschungszentrum Karlsruhe, FZKA 7491, (2009).

[141] H. Schubert, *Anisotropic Thermal Expansion Coefficients of Y_2O_3-Stabilized Tetragonal Zirconia*, Journal of the American Ceramic Society, 69, 3 (1986), pp. 270–271.

[142] T. J. Seebeck, *Magnetische Polarisation der Metalle und Erze durch Temperatur-Differenz. (1822-1823)*, A. J. v. Oettingen, Verlag von Wilhelm Engelmann, (1895).

[143] S. Shafiee and E. Topal, *An econometrics view of worldwide fossil fuel consumption and the role of US*, Energy Policy, 36 (2008), pp. 775–786.

[144] S. Shin, H. N. Cho, B. S. Kim, and H. H. Cho, *Influence of upper layer on measuring thermal conductivity of multilayer thin films using differential 3-ω method*, Thin Solid Films, 517 (2008), pp. 933–936.

[145] G. J. Snyder and E. S. Toberer, *Complex thermoelectric materials*, Nature Materials, 7 (2008), pp. 105–114.

[146] Q. Song, Z. Cui, S. Xia, Z. Chen, and J. Zhang, *Measurement of SiN_x thin film thermal property with suspended membrane structure*, Sensors and Actuators A: Physical, 112 (2004), pp. 122–126.

[147] X. Song, M. Xie, F. Zhou, G. Jia, X. Hao, and S. An, *High-temperature thermal properties of yttria fully stabilized zirconia ceramics*, Journal of Rare Earths, 29 (2011), pp. 155–159.

[148] A. Stranz, U. Sökmen, J. Kähler, A. Waag, and E. Peiner, *Measurements of thermoelectric properties of silicon pillars*, Sensors and Actuators A: Physical, 171 (2011), pp. 48–53.

[149] T. R. Taylor, P. J. Hansen, B. Acikel, N. Pervez, R. A. York, S. K. Streiffer, and J. S. Speck, *Impact of thermal strain on the dielectric constant of sputtered barium strontium titanate thin films*, Applied Physics Letters, 80 (2002), pp. 1978–1980.

[150] THERMOELECTRIC. *http://www.thermoelectric.com/2005/images/te.jpg*, (11.20 a.m., 09. Aug. 2012).

[151] R. J. D. Tilley, *Understanding Soilds - The Science of Materials*, John Wiley and Sons, Inc., (2004).

[152] T. Tong and A. Majumdar, *Reexamining the 3-omega technique for thin film thermal characterization*, Review of Scientific Instruments, 77 (2006), pp. 104902–9.

[153] TUVIE. *http://www.tuvie.com/wp-content/uploads/Orange- power- whellies2.jpg*, (09:54 a.m., 23 Jan. 2013).

[154] E. Tyulyukovskiy and N. Huber, *Identificationof visco plastic material parameters from spherical identation data. Part I: Neural networks*, Journal of Materials Research, 21 (2006), pp. 664–676.

[155] J. van den Brink and D. I. Khomskii, *Multiferroicity due to charge ordering*, Journal of Physics: Condensed Matter, 20 (2008), p. 434217.

[156] R. Venkatasubramanian, *Lattice thermal conductivity reduction and phonon localizationlike behavior in superlattice structures*, Physical Review B, 61 (2000), pp. 3091–3097.

[157] R. Venkatasubramanian, E. Siivola, T. Colpitts, and B. O'Quinn, *Thin-film thermoelectric devices with high room-temperature figures of merit*, Nature, 413 (2001), pp. 597–602.

[158] C. B. Vining, *An inconvenient truth about thermoelectrics*, Nature Materials, 8 (2009), pp. 83–85.

[159] F. Völklein, *Thermal conductivity and diffusivity of a thin film $SiO_2 - Si_3N_4$ sandwich system*, Thin Solid Films, 188 (1990), pp. 27–33.

[160] A. Wahl, V. Hardy, C. Martin, and C. Simon, *Magnetic contributions to the low-temperature specific heat of the ferromagnetic insulator $Pr_{0.8}Ca_{0.2}MnO_3$*, The European Physical Journal B, 26 (2002), pp. 135–140.

[161] H. Wang and M. Sen, *Analysis of the 3-omega method for thermal conductivity measurement*, International Journal of Heat and Mass Transfer, 52 (2009), pp. 2102–2109.

[162] X. Wang and P. Xiao, *Residual stresses and constrained sintering of YSZ/Al_2O_3 composite coatings*, Acta Materialia, 52 (2004), pp. 2591–2603.

[163] Z. L. Wang, D. W. Tang, and X. H. Zheng, *Simultaneous determination of thermal conductivities of thin film and substrate by extending 3ω-method to wide-frequency range*, Applied Surface Science, 253 (2007), pp. 9024–9029.

[164] S. Wiedigen, T. Kramer, M. Feuchter, I. Knorr, N. Nee, J. Hoffmann, M. Kamlah, C. A. Volkert, and C. Jooss, *Interplay of point defects, biaxial strain, and thermal conductivity in homoepitaxial srtio₃ thin films*, Applied Physics Letters, 100 (2012), pp. 061904–4.

[165] W. Xu, J. Li, G. Zhang, X. Chen, R. Galos, H. Hadim, M. Lu, and Y. Shi, *A low-cost MEMS tester for measuring single nanostructure's thermal conductivity*, Sensors and Actuators A: Physical, 191 (2013), pp. 89–98.

[166] Y. Xu, Z. Shi, C. Guo, and J. Zhao, *Analysis of key stresses of coated tool in high speed cutting*, Applied Mechanics and Materials, 151 (2012), pp. 442–445.

[167] J. Yu, Z. Tang, F. Zhang, H. Ding, and Z. Huang, *Measurement of the Heat Capacity of Copper Thin Films Using a Micropulse Calorimeter*, Journal of Heat Transfer, 132 (2010), pp. 012403–012406.

[168] J.-K. Yu, S. Mitrovic, D. Tham, J. Varghese, and J. R. Heath, *Reduction of thermal conductivity in phononic nanomesh structures*, Nature Nanotechnology, 5 (2010), pp. 718–721.

[169] E. Yusibani, P. Woodfield, M. Fujii, K. Shinzato, X. Zhang, and Y. Takata, *Application of the Three-Omega Method to Measurement of Thermal Conductivity and Thermal Diffusivity of Hydrogen Gas*, International Journal of Thermophysics, 30 (2009), pp. 397–415.

[170] J. Zhang, A. A. Heitmann, S. P. Alpay, and G. A. Rossetti Jr., *Electrothermal properties of perovskite ferroelectric films*, Journal of Material Science, 44 (2009), pp. 5263–5273.

[171] K. L. ZHANG, S. K. CHOU, AND S. S. ANG, *Performance Prediction of a Novel Solid-Propellant Microthruster*, Journal of Propulsion and Power, 22, 1 (2006).

[172] B. ZHENG, Z. XIAO, B. CHHAY, R. ZIMMERMAN, M. E. EDWARDS, AND D. ILA, *Thermoelectric properties of MeV Si ion bombarded Bi_2Te_3/Sb_2Te_3 superlattice deposited by magnetron sputtering*, Surface and Coatings Technology, 203 (2009), pp. 2682–2686.

[173] Z.-X. ZONG, Z.-J. QIU, S.-L. ZHANG, R. STREITER, AND R. LIU, *A generalized 3ω method for extraction of thermal conductivity in thin films*, Journal of Applied Physics, 109 (2011), pp. 063502–8.

Acknowledgment - Danksagung

Die vorliegende Arbeit wurde im Teilinstitut Werkstoff- und
Biomechanik (WBM) des Instituts für Angewandte Materialien (IAM)
am Karlsruher Institut für Technologie (KIT) im Zeitraum von
Oktober 2009 bis Mai 2014 angefertigt.

Zu Beginn möchte ich meinem Betreuer Herrn Prof. Dr. Marc
Kamlah außerordentlich danken. Nur durch seine fachlich sehr
kompetente Betreuung und zahlreichen Hilfestellungen bei
multiphysikalischen Fragen war es mir möglich das benötigte
Verständnis zu entwickeln und diese Arbeit anzufertigen. Ich danke
ihm auch sehr für ein freundschaftliches Arbeitsklima, bei dem man
sich unter viel Freiraum trotzdem stets gefordert fühlt.

Meinen Kooperations- und Projektpartnern aus Clausthal und
Göttingen danke ich für viele interessante Arbeitstreffen. Vor allem
den Arbeitsgruppen von Prof. Dr. Christian Jooss und Prof. Dr.
Cynthia Volkert danke ich für die vielen Fragestellungen und
Anregungen. Im Besonderen möchte ich auch Stefanie Wiedigen,
Thilo Kramer und Malte Scherff für die experimentellen Messungen
dieser Arbeit sowie hilfreichen Diskussionen danken. Darüber
hinaus geht ein spezieller Dank an Herrn Prof. Dr. Völklein. Er beriet
mich nicht nur mit jahrelangem Fachwissen zu Membransystemen
sondern stellte mir auch thermische und mechanische Materialdaten
zur Verfügung.

Ebenso möchte ich Prof. Kamlah für die Übernahme des
Hauptreferats sowie Prof. Jooss für die Übernahme des ersten
Korreferats danken. Herrn Prof. Dr. Oliver Kraft danke ich für die
Aufnahme an seinem Institut und die Übernahme des zweiten
Korreferats.

Auch bei Dr. Ingo Münch möchte ich mich bedanken. Er wecke bei
mir das Interesse für wissenschaftliches Arbeiten während meiner
Diplomarbeit und führte mich in das interessante Feld der
multiphysikalischen Simulationen ein. Ebenso gilt mein Dank Herrn
Prof. Dr. Werner Wagner. Durch seine Studiums begleitende
Betreuung als Mentor war es mir möglich einen englischsprachigen
Auslandsaufenthalt zu absolvieren und dabei vor allem meine
Englischkenntnisse zu erwerben.

Ich danke allen Kollegen für zahlreiche fachliche Diskussionen und
eine wunderbare Arbeitsatmosphäre in den letzten Jahren. So möchte
ich auch im Speziellen der Arbeitsgruppe von Prof. Kamlah, sowie
meinen langjährigen Zimmerkollegen danken. Darüber hinaus
bedanke ich mich bei Frau Dr. Svetlana Scherrer-Rudiy für die
Einührung in die Parameteridentifikation mittels Neuronaler Netze.
Ebenso gilt mein Dank Matthew Berwind für das Lektorieren dieser
in Englisch geschriebenen Arbeit.

Am Ende möchte ich mich ganz herzlich bei meinen Eltern Angelika
und Klaus, meinem Bruder Julius, sowie meiner Freundin Evelyn, für
jegliche Unterstützung und Motivation auf meinem bisherigen
Lebensweg bedanken.

Manuel Feuchter, Frühjahr 2014

Schriftenreihe
des Instituts für Angewandte Materialien

ISSN 2192-9963

Die Bände sind unter www.ksp.kit.edu als PDF frei verfügbar
oder als Druckausgabe bestellbar.

Band 1   Prachai Norajitra
          Divertor Development for a Future Fusion Power Plant. 2011
          ISBN 978-3-86644-738-7

Band 2   Jürgen Prokop
          **Entwicklung von Spritzgießsonderverfahren zur Herstellung
          von Mikrobauteilen durch galvanische Replikation.** 2011
          ISBN 978-3-86644-755-4

Band 3   Theo Fett
          **New contributions to R-curves and bridging stresses –
          Applications of weight functions.** 2012
          ISBN 978-3-86644-836-0

Band 4   Jérôme Acker
          **Einfluss des Alkali/Niob-Verhältnisses und der Kupferdotierung
          auf das Sinterverhalten, die Strukturbildung und die Mikro-
          struktur von bleifreier Piezokeramik $(K_{0,5}Na_{0,5})NbO_3$.** 2012
          ISBN 978-3-86644-867-4

Band 5   Holger Schwaab
          **Nichtlineare Modellierung von Ferroelektrika unter
          Berücksichtigung der elektrischen Leitfähigkeit.** 2012
          ISBN 978-3-86644-869-8

Band 6   Christian Dethloff
          **Modeling of Helium Bubble Nucleation and Growth
          in Neutron Irradiated RAFM Steels.** 2012
          ISBN 978-3-86644-901-5

Band 7   Jens Reiser
          **Duktilisierung von Wolfram. Synthese, Analyse und
          Charakterisierung von Wolframlaminaten aus Wolframfolie.** 2012
          ISBN 978-3-86644-902-2

Band 8   Andreas Sedlmayr
          **Experimental Investigations of Deformation Pathways
          in Nanowires.** 2012
          ISBN 978-3-86644-905-3

Band 39   Jörg Ettrich
           Fluid Flow and Heat Transfer in Cellular Solids. 2014
           ISBN 978-3-7315-0241-8

Band 40   Melanie Syha
           **Microstructure evolution in strontium titanate Investigated by
           means of grain growth simulations and x-ray diffraction contrast
           tomography experiments.** 2014
           ISBN 978-3-7315-0242-5

Band 41   Thomas Haas
           **Mechanische Zuverlässigkeit von gedruckten und gasförmig
           abgeschiedenen Schichten auf flexiblem Substrat.** 2014
           ISBN 978-3-7315-0250-0

Band 42   Aron Kneer
           **Numerische Untersuchung des Wärmeübertragungsverhaltens
           in unterschiedlichen porösen Medien.** 2014
           ISBN 978-3-7315-0252-4

Band 43   Manuel Feuchter
           **Investigations on Joule heating applications by multiphysical
           continuum simulations in nanoscale systems.** 2014
           ISBN 978-3-7315-0261-6